Introduction

General Electric and the Grand Experiment is about the illegal and immoral dumping of polychlorinated biphenyls (PCBs) in the rivers , streams and land across the northeastern United States ; and their continued flaunting of environmental rules and regulations .

I have also attempted to look at what might be done to mitigate some of the damage done by looking at the latest research into some of the diseases that were caused by this dumping ; type 1 diabetes in particular .

Peace

Fred Taikowski

Cover Image Explanation :

This is an image of the Schenectady, NY General Electric plant (taken in 1896) .

It was from this location that huge amounts of polychlorinated biphenyls were dumped directly into the Hudson River .

(see ' Image Credits ' at end of book for usage licenses)

Dedication

This book is dedicated to my grandfather , Joseph Taikowski Sr. ; who worked in the General Electric transformers department for 35 years ;

And also to my father ; Joseph Taikowski , Jr. ; who died much too young from complications from type 1 diabetes .

Chapter 01

It was 1888 .

A young gentleman named Cummings C. Chesney was on the laboratory staff of one William Stanley Jr. in Great Barrington, Massachusetts . In 1890 , Mr. Chesney and Mr. Stanley moved to Pittsfield , MA , and they organized the Stanley Electrical Manufacturing Company .

In 1902 this Company was purchased by the General Electric Company (1)

It was January of 1891 when this company first began to manufacture electrical transformers on Clapp Avenue (now gone) . (2) After changing hands two times , the main Morningside plant of the General Electric employed 1200 people by the year 1901 . By that time both of the other locations at Clapp Avenue and Renne Avenue were closed .

The Morningsdie location continued it's expansion ; until by 1907 the sprawling GE plant covered 1,600,000 square feet of buildings . In 1915 one-sixth of the population of Pittsfield were employed at the plant .

At first, only small standard transformers (known as Type 'H') were built in Pittsfield . By 1908 , however , when the Lynn , MA and the Schenectady , NY departments were moved to Pittsfield , all GE transformers were being manufactured in Pittsfield , with the exception of a few specialty types . (3) The plant had also expanded in size by another 50 percent .

The question that we must now ask ourselves is : What , exactly , were these transformers , and why were they so important ?

Let us look at , then , the basic transformer that was being built at the GE in Pittsfield from 1908 through 1932 . At this time the United States (and the world , for that matter) was just entering the electric age . Demand was growing in the burgeoning factories and in the ever expanding cities for electrical power . And General Electric , and William Stanley was in competition with several others to see just who was going to supply those needs .

A definition of a basic transformer is in order here : " A transformer is an electrical device that transfers energy between two or more circuits through electromagnetic induction. " (4) . And as these transformers grew larger and larger to meet the expanding energy needs , they needed to be cooled .

Why would they need to be cooled ? Well , a simple rule for transformer cooling is as follows : " the temperature is halved for about every 7 °C to 10 °C increase in operating temperature " (5)

To continue : " Small dry-type and liquid-immersed transformers are often self-cooled by natural convection and radiation heat dissipation. As power ratings increase, transformers are often cooled by forced-air cooling, forced-oil cooling, water-cooling, or combinations of these. Large transformers are filled with transformer oil that both cools and insulates the windings " (5)

And one of these ' transformers oils ' , in the larger transformers at the General Electric , was the chemical compound Polychlorinated biphenyl , or ' PCB ' for short . The first PCB chemical compounds had been developed in German laboratories in the year 1881 . (6) Why were PCBs used instead of less hazardous chemical compunds ? Because the PCBs were found to be non - flammable , which is very important in their use to cool large electronic components .

So the production of PCBs was taken over by the Monsanto Chemical Company in 1929 , and introduced into large scale use at Building 1 of the General Electric complex in 1932 . (6)

But - - there were problems . Almost immediately .

(1) Cummings C. Chesney ; IEEE Global History Network ;
http://www.ieeeghn.org/wiki/index.php/Cummings_C._Chesney

(2) Archives:Transformers at Pittsfield, part 1 : IEEE Global History Network ;
http://www.ieeeghn.org/wiki/index.php/Archives:Transformers_at_Pittsfield%2C_
part_1#Chapter_1:_The_Early_Years

(3) Chapter 1: The Early Years ; IEEE Global History Network ;
http://www.ieeeghn.org/wiki/index.php/Archives:Transformers_at_Pittsfield%2C_
part_1#Chapter_1:_The_Early_Years

(4)(5) Transformers : Wilipedia   :
http://en.wikipedia.org/wiki/Transformer

(6) Polychlorinated bipheny : History :
http://en.wikipedia.org/wiki/Polychlorinated_biphenyl#Massachusetts

Chapter Two

Transformers Types and Sizes

plus PCB Cooling

When we are talking about transformer production at the General Electric in the 1920s , we are not talking about the small size transformer such as the laminated core , or the toroidal . No , what we are talking about here are the large oil cooled transformer which could be as large as 10 feet by 10 feet ; all produced in a large interior room . Quite a large percentage of this type of transformer (the type that was capable of producing enough electricity to supply the needs of the growing cities) , a shipment of several transformers , each capable of producing up to 22,500 to 55,000 volts , rolled steadily onto the traincars which pulled up at GE Pittsfield ; for shipment around the world . (1)

Business seemed to be good for the GE in the 1920s . The use of a two - phase wiring system had allowed their transformers to produce electricity at a 41 % higher rate than their competitors : and thus they were in high demand . Production of these units were thusly expanded from Building 1 at the Pittsfield plant to other buildings within the compound ; and substations were installed within the plant to step down the elecctrical power for use in the plant .

Yes , all seemed to going well . Until .... reports began to filter in about the dangers of using the oil compound PCB that was used to cool these large transformers . In 1936 one Dr. Lewis Schwartz, Senior Surgeon with the United States Public Health Service , wrote a paper in which he warned of the dangers of Pyrenol, the type of PCBs used by GE. (2) In his paper Dr. Schwartz wrote :

 " "In addition to the skin lesions, symptoms of systematic poisoning have occurred among workers inhaling the fumes. Those working with the chloro diphenyls (PCBs) have complained of digestive disturbances, burning of the eyes, impotence and hematuria. The latter symptom developed among a number of men making amino diphenyl, which is used in the making of a rubber antioxidant. Causes of death from yellow atrophy of the liver have been reported among workers exposed to the fumes of the chloro naphthalenes." (2)

A short time later Dr . Schwartz wrote : " Also in 1936, Dr. Schwartz cautioned in an article that "workers in chlorinated naphthalenes and di phenyls (PCBs) should be periodically examined for symptoms of systemic poisoning." (2) And GE and it's management team was quite aware of the risked involved with using Polychlorinated biphenyl (PSBs ; also called halowax by one it's manufacturers .

An Assistant GE Plant manager in Pennsylvania , F. R. Kaimer, had this to say about the PCB situation :

"It is only one and a half years ago that we had in the neighborhood of 50 to 60 men afflicted with various degrees of this acne about which you all know. Eight or ten of them were very severely afflicted – horrible specimens as far as their skin conditions were concerned. One man died and the diagnosis may have attributed his death to exposure to halowax vapors but we are not sure of that ... we had 50 other men in very bad condition as far as the acne was concerned. The first reaction that several of our executives had was to throw it out – get it out of our plant. They didn't want anything like that for treating wire. But that was easily said but not so easily done. We might just as well have thrown our business to the four winds and said, 'We'll close up,' because there was no substitute and there is none today in spite of all the efforts we have made through our own research laboratories to find one." (2)

So , despite efforts being made by GE laboratories there was no success in finding a substitute for PCBs to cool their all important transformers . Thus , their use of it continued . PCBs were used through the 1970s , when it was phased out gradually after it's use was banned . This , of course , does not mean that the danger of PCBs automatically ' disappeared ' when it's use was discontinued in the 1970s . Not by a long shot . Now came the problem of the clean- up , because Pittsfield and the surrounding area , with it's lakes and rivers , had been polluted , and was saturated with the chemical .

How did this happen ? How did Polychlorinated biphenyl get out of the confines of the GE plant , and into the groundwater and land surrounding the plant ?

The most obvious answer is from spills and leakage from the transformers themselves . How exactly was this accomplished ?

On this point let us turn to a quotation from one Ed Bates, former Manager of Tests at the Power Transformer division at GE in Pittsfield on the subject of Pyranol, GE's name for its version of PCBs :

"We used to use an average of 20,000 gallons of Pyranol a week and this is if you do simple mathematics, this is one hundred and forty thousand pounds of ... PCBs a week that we were handling. And we had a loss rate: spillage, overfilling, of about 3% so this says that every week we would lose between four and five thousand pounds of PCBs that would go down into the drain and into the river. ...About a million and a half pounds of PCBs have been plowed into that river. I imagine a good 30% is left." (3)

So we are talking here four to five thousand pounds of PCBs that would go directly into the Housatonic River (from 1932 through 1977 , the time period that this website , the Berkshire Environmental Action Team , states that PCBs were dumped . Of course the Houstonic River was not the only site in the city to receive these polluting PCBs . A large amount of PCBs entered into Silver Lake , a lake which was adjacent to the large General Electric complex on East Street in Pittsfield   (and connected to the Housatonic River) . When I was a resident of Pittsfield (I am originally from the Pittsfield , Mass area ; and lived on and off there for approximately 25 years . I now reside in Orlando , FL .) . But , at any rate , when I was resident of Pittsfield , Silver lake was known as ' the most polluted lake in the world ' . And it looked the part , too . This was an absolutely black cesspool that was located near the downtown area of the city . It was only after GE applied for a permit to dump PCBs in 1975 that the environmental community sat up , and took notice ; and it is just now , and it has just been recently , some 35 years after the dumping of PCBs was ordered to be halted in 1979 , that Silver Lake has begun a comeback that once again allows it to support fish and wildlife .

And it is not , of course , just the Housatonic River watershed . (The Housatonic River is a large river which runs the length of Berkshire County .) The General Electric had the gall to try and dispose of it's PCB waste material by offering it , or PCB tainted soils , as landfill , or fill for residents of Berkshire County . The Newell Street area of Pittsfield was a particularly polluted area , According to a September 2003 report by the U.S. Department of Health and Human Services , the amount of PCB contaminated soil here ranged form : " ....655 parts per million (ppm) and range as high as 25,500 ppm in some hot spots on the site. " (4)   According to this report , however , this Newell Street area site was not considered extremely hazardous due to the fact that it had high fences around it , and quite a bit of growth that prohibited the public from coming into contact with it . This was 2003 , of course , and I am not privy to what exactly is going on in this area as of 2015 .

   The Newell Street area was , of course , not the only site in the city to have contaminated soil . According to this 1997 report from the EPA " Primarily in the 1940s and 1950s , trucks delivered fill from the GE facility (often at the request of the property owner) to many low-lying marshy areas in and around Pittsfield . Some of these marshy areas are former oxbows . Other areas are just low-lying or uneven properties where the owner asked GE tp deliver fill so the property would be level . Some , but not all , of the fill that GE provided was contaminated with PCBs . PCBs are bound lightly to the fill , and do not move from the fill into groundwater , or into other clean areas of soil or fill . Fill also may have been received from sources other than GE . Non GE fill may or may not be contaminated . " (5)

   I am , it by now must be fairly evident , concentrating on pollution in the Pittsfield , Mass . area . This is because I am a former resident , and my grandfather (Joseph Taikowski , Sr .) worked for GE Transformers for 35 yeras . I also inherited a house (next to my grandfather's) just off of East Street in Pittsfield ; which was just across the street from GE Plastics (on Plastics Avenue) , and just down the street from Silver Lake and Building 1 (GE Transformers) .
   No , not by any means was Pittsfield the only polluted area here in the United States . The pollution of the Hudson River in New York State , downstream from the Rensselaer , NY General Electric is a very famous case . The Anniston , Alabama case is perhaps the largest case in the U.S. Here Monsanto , the same company that had supplied PCBs for the General Electric in Pittsfield , was found directly responsible for dumping PCBs in Anniston's creeks , and poured millions of pounds of the chemical into open pits . The resultant health problems were enormous . In 2003 Monsanto was fined $ 750 million dollars for it's actions in Anniston .
   Industrialized parts of the midwest , such as Ohio and Michigan , were also subject to such pollution , where General Electric , Monsanto and their subsidiary companies had locations .

   And , of course , as General Electric has moved overseas , in order to take advantage of cheap labor and lax environmental regulations , the environment suffers with each stop . And at each location the diseases caused by PCB poisoning multiply .

(1) Schenectady Electrical Handbook : The Schenectady Works of the General Electric Company :
http://www.schenectadyhistory.org/resources/seh/transformers.html

(2) Berkshire Environmental Action Team : Did GE Know About The Dangers Of PCBs? : http://www.thebeatnews.org/BeatTeam/ge-knew/

(3) Berkshire Environmental Action Team : General Electric and PCBs   :
http://www.thebeatnews.org/BeatTeam/ge-pcbs/

(4) U.S. Department of Health and Human Services : Public Health Assessment for General Electric Site Newell Street Area II :
http://www.mass.gov/eohhs/docs/dph/environmental/investigations/general-electric/ge-newell2.pdf

(5) Polychlorinated Bipheyls (PCBs) : A Fact Sheet : How did PCBs get from the GE into the environment ? :
http://www.epa.gov/region1/ge/pcbshealthandenviro/pcbfact.pdf

Chapter Three

Further Facts on PCBs

In this chapter I shall try and explain exactly how the Polychlorinated biphenyl dioxin is able to damage the human cell , causing a myriad of diseases .

Most of my own research has concentrated on diabetes and PCBs (as I am a severe Type 1 diabetic , and have been for the past 52 years . However , PCB exposure has been linked to cancer , and , according to the EPA website , " EPA has found clear evidence that PCBs have significant toxic effects in animals, including effects on the immune system, the reproductive system, the nervous system and the endocrine system. " (1)

The Hudson River PCB Story describes PCB's as : " The International Agency for Research on Cancer and the Environmental Protection Agency classify PCBs as a probable human carcinogen. The National Toxicology Program has concluded that PCBs are reasonably likely to cause cancer in humans. The National Institute for Occupational Safety and Health has determined that PCBs are a potential occupational carcinogen. Studies of PCBs in humans have found increased rates of melanomas, liver cancer, gall bladder cancer, biliary tract cancer, gastrointestinal tract cancer, and brain cancer, and may be linked to breast cancer. PCBs are known to cause a variety of types of cancer in rats, mice, and other study animals. " [2]

This particular report (the Hudson River PCB Story) goes on to report it's findings that PCB's can cause : developmental effects , disruption of hormone function and immune system and thyroid effects . (2)

The website ' Green Facts ' describes health effects of PCB's thusly :

" Nonetheless, the evidence suggests that exposure to PCBs is associated with an increased risk of certain cancers of the digestive tract, liver and skin. PCB exposure is also associated with reproductive deficiencies, such as reduced growth rates, retarded development, and certain neurological effects which may or may not persist beyond infancy. The immune system can also be affected, leading to increased infection rates, and changes in the skin such as chloracne and pigmentation disturbances. " (3)

So , there are many different sources that provide us with a horror story of PCB exposure effects . The question I would like to address is just how the chemicals in PCBs react in order to do their damage .

Before I do this , I must mention the extent that spilled PCBs remain in the environment . As a general rule the more chlorine atoms that are present in the PCB molecule (there are different types of PCBs that are produced) the longer it takes for the PCB to break down , and the longer it remains in the atmosphere , in water , and soil . As a general rule it takes PCBs 17 to 210 days for half of the amount (initially) present to be broken down in shallow water for molecules with 1 to 4 chlorine atoms. (4) In the atmosphere " it takes for half of the amount of PCBs (initially) present to be broken down ranges from 3.5 to 83 days for molecules with 1 to 5 chlorine atoms. (4) In the soil there is no specific time table available , because there are too many different factors involved : the number and location of chlorine atoms, PCB concentration, the type of microorganisms present, available nutrients, and temperature. (4)

So PCBs persist in the air , water and soil ; and all that it takes is coming into contact with the chemical over a period of time to suffer from PCB infection .

First an explanation of exactly what PCBs consist of . PCBs are chemicals formed by attaching one or more chlorine atoms to a pair of connected benzene rings. (5)

" Depending on the number and position of chlorine atoms attached to the biphenyl ring structure, 209 different PCB congeners can be formed. PCB congeners can be divided into the coplanar, the mono-ortho-substituted PCBs, and other non-dioxin-like PCBs. The significance of this designation is that coplanar and some of the mono-ortho-substituted PCBs have dioxin-like toxicologic effects." (5) The chlorination pattern of the PCB molecule determines it's toxicity .

Between 1929 and 1977 more than 1.25 billion pounds of PCB were produced in the United States . If there are no extenuating circumstances (as listed above) PCBs can exist in the soil from months to years . (5) The safety level for PCB exposure for air in the workplace is 1.0 mg / million for PCBs with 42% Chlorine ; and 0.5 mg/million for PCBs with 54% Chlorine . (5)

Here is the EPA safety levels :

Drinking water : environment   :
0.0005 ppm

Enforceable Tolerance level for Food (in the environment) :
0.2 to 3.0 ppm (parts per million)

Paper Food Packaging   :   10
ppm

Allowable Daily Food intake level   :
6.0 ppm

  And How does the PCB actually ' do it's thing ' ? (Cause damage to the human being that is exposed to it ?) Here is a rather complex explanation from the ATSDR Case Studies that I will attempt to simplify :

  (Note : the ATSDR is the Agency for Toxic Substances and Diseases Registry . Their website is www.atsdr.cdc.gov)

  (5) PCBs are metabolized by the microsomal monooxygenase system catalyzed by cytochrome P-450 to phenols (via arene oxide intermediates), which can be conjugated or further hydroxylated to form a catechol. Arene oxide intermediates are electrophilic in nature. They can covalently bind to nucleophilic cellular macromolecules (e.g., protein, DNA, RNA) and induce DNA strand breaks and DNA repair, which can contribute to the toxic response of PCBs. Additionally, arene oxide intermediates can be conjugated with glutathione and further metabolized to form methylsulfonyl metabolites, which have been identified in human serum and tissue samples and in laboratory animals.

Binding of methylsulfonyl metabolites to some proteins may contribute to some of the toxic effects of PCBs. It has also been hypothesized that hydroxylated PCB metabolites could contribute to the toxicity of PCBs . " (5)

So PCBs cause DNA strand breaks , and hinders DNA repair . PCBs also metabolize within the human to form methylsulfonyl metabolites , which , when binding to proteins adds to the toxicity effect of PCBs .

Furthermore , in the ATSDR study , there can be damage to the Endocrine system . Here is their statement : " Limited but corroborative occupational data indicate a potential for toxic effects in the thyroid system in humans." (5) Thyroid hormones are essential for normal behavioral, intellectual, and neurologic development. (5) They further state :

" Recent studies in populations exposed to PCBs and chlorinated pesticides found a dose-dependent elevated risk of diabetes . " This report was written in 2008 .

The most dramatic findings in this study was in the field of cancer research . This study states : " A retrospective analysis of a study of two plants that manufactured electrical capacitors in the United States found a significant increase in the incidence of cancer. The primary target tissues for the cancers were the liver, gallbladder, and biliary tract ." (5)

" Likewise, an increased incidence of melanomas associated with exposure to PCBs has also been observed for workers who manufactured capacitors [Bahn et al. 1976; Ruder et al. 2006; Sinks et al. 1992]. Sinks et al. [1992] observed the increased risks for brain cancer among workers exposed to PCBs in an electrical capacitor manufacturing plant in Indiana, and this finding has been further confirmed by a recent study from Ruder et al. [2006]. (5)

Further on the cancer studies :

" A recent analysis of a cohort of 24,865 capacitor-manufacturing workers exposed to PCBs at three plants showed evidence of associations between cumulative exposure to PCBs and increased total cancer and intestinal cancer mortality among female long-term workers and excess myeloma for male long-term workers [Ruder et al. 2014]. " (5)

Different mixtures of PCBs had different potencies and, thus, different toxicity , but after this study it was concluded : " On the basis of these laboratory data, EPA has determined that PCBs are probable human carcinogens and has assigned them the cancer weight-of-evidence classification B2 [IRIS 2012].

At present there are " no specific treatment exists for PCB accumulation. Patients should avoid further PCB exposure and also avoid other hepatotoxic substances, including ethanol." (5) There is no medication at present to remove or flush PCBs out of the system . The only thing that now exists is to treat the symptoms once they present themselves .

Symptoms of Chronic exposure are as follows :

• Abdominal pain,
• Anorexia,
• Jaundice,
• Nausea,
• Vomiting,
• Weight loss
• Uroporphyria.

Headache, dizziness, and edema have also been reported . (5)

Technically Polychlorinated biphenyl is " a synthetic organic chemical compound of chlorine attached to biphenyl, which is a molecule composed of two benzene rings. There are 209 configurations of organochlorides with 1 to 10 chlorine atoms. The chemical formula for a PCB is $C_{12}H_{10-x}Cl_x$. 130 of the different PCB arrangements and orientations are used commercially. (6)

Simply put , it is different combinations of chlorine molecules attached to two rings of benzine (benzine is a major component in the production of gasoline) .

Because there are some many different configurations of the PCB compound , it's effects when inhaled , ingested or absorbed will vary greatly , mainly depending upon the number of chlorine molecules that the compound contains .

So , once it is absorbed , how does the compound do it's damage ?

In this regard I have found two theories .

The first is proposed on the Wikipedia description of dioxins , and dioxin - like compounds . (See source below (7)
It reads as follows :

" The toxicity is mediated through the interaction with a specific intracellular protein, the aryl hydrocarbon (AH) receptor, a transcriptional enhancer, affecting a number of other regulatory proteins "
" TCDD binding to the AH receptor induces the cytochrome P450 1A class of enzymes which function to break down toxic compounds, e.g., carcinogenic polycyclic hydrocarbons such as benzo(a)pyrene "

So , this induces the cytochrome P450 1A class of enzymes which functions to break down toxic compounds . I believe (as far as diabetes is concerned) that it is this enzyme that may destroy the insulin producing cells . The above conclusions were reached from the following studies :

1) Dencker L (1985). "The role of receptors in 2,3,7,8-tetrachlorodibenzo-p-dioxin (TCDD) toxicity

2) L. Poellinger. Mechanistic aspects – the dioxin (aryl hydrocarbon) receptor (2000). "Mechanistic aspects—the dioxin (aryl hydrocarbon) receptor.". Food Additives and Contaminants

3) J. Lindén, S. Lensu, J. Tuomisto, R. Pohjanvirta. (2010). "Dioxins, the aryl hydrocarbon receptor and the central regulation of energy balance. A review.". Frontiers in Neuroendocrinology

4) Okey, A. B. (2007). "An aryl hydrocarbon receptor odyssey to the shores of toxicology: The Deichmann lecture". International congress of toxicology-XI. Toxicological Sciences

A differing argument , that stated by ' Diabetes and Environment ' (See Source below (8) states :

" Dioxin has been shown to stimulate insulin secretion by rat beta cells (Kim et al. 2009). The authors suggest that dioxin may therefore contribute to the risk of developing diabetes by causing continuous insulin release, followed by beta cell dysfunction. Other studies have found that dioxin exposure impaired insulin secretion from rodent beta cells (Kurita et al. 2009, Novelli et al. 2005), and, at higher doses, even killed them (Piaggi et al. 2007). Hectors et al. (2011) review the effects of chemicals on beta cells, and find that other chemicals have also been found to increase as well as decrease insulin secretion. The effect may depend on dose, the animal or cells used in the experiment, or other factors. Dioxin's ability to affect beta cells may have importance for diabetes development."

So this study is stating that dioxins are capable of over - stimulating insulin production in the beta cells ; thus causing them to fail . This study , as far as I am concerned , is somewhat inconclusive (and it has been done only in rodents) . So , I will focus mostly on the first set of studies (those quoted in Wikipedia) .

Once the dioxin PCB (or any of it's toxic configurations) enter the human body at levels higher than I have quoted above (and in the Chart on the following page)   the damage begins when the PCB compound starts to accumulate . Below is the statement by Green Facts on how PCB's are absorbed into the body :

" PCBs can enter human cells and tissues when contaminated air is breathed in, when contaminated food enters the digestive system, or through contact with the skin. Tests on laboratory animals show that PCBs are readily absorbed through the digestive tract when swallowed, and to a lesser extent through the skin. The main PCB elimination routes are through the faeces, urine, and breast milk.
Once in the gastrointestinal tract, ingested PCBs diffuse across cell membranes and enter blood vessels and the lymphatic system. PCBs, especially those that contain a greater number of chlorine atoms, are readily soluble in fats and thus tend to accumulate in fat-rich tissues such as the liver, brain and skin. " (9)

So these are the situations that occur when man - made PCBs are created , and then absorbed into the human system in one fashion or another . But , what about diabetes before the introduction of man - made PCB's in 1881 ? Are there any relationships in the two conditions ? Does PCB (or something similar) occur naturally ? I shall attempt to answer these questions in the next chapter .

(1) U.S. Environmental Protection Agency : Polychlorinated Biphenyls :
Health Effects of PCBs :
http://www.epa.gov/epawaste/hazard/tsd/pcbs/pubs/effects.htm

(2) Clearwater Presents The Hudson River PCB Story : What Are The
Human Health Effects Of PCBs? :
http://www.clearwater.org/news/pcbhealth.html

(3) Green Facts : PCBs Polychlorinated biphenyls :
http://www.greenfacts.org/en/pcbs/l-2/6-effects-human.htm

(4) Green Facts : To what extent do PCBs break down or persist
in the environment? : http://www.greenfacts.org/en/pcbs/l-2/2-
biomagnification.htm

(5) ATSDR Case Studies in Environmental Medicine Polychlorinated
Biphenyls (PCBs) Toxicity :
http://www.atsdr.cdc.gov/csem/pcb/docs/pcb.pdf

(6) Wikipedia : Polychlorinated biphenyl :
http://en.wikipedia.org/wiki/Polychlorinated_biphenyl

(7) Wikipedia : Dioxins and dioxin-like compounds :
http://en.wikipedia.org/wiki/Dioxins_and_dioxin-like_compounds

(8) Diabetes and the Environment : Dioxin :
http://www.diabetesandenvironment.org/home/contam/pops/dioxins

(9) Green Facts : What happens to PCBs when they enter the
body? : http://www.greenfacts.org/en/pcbs/l-2/4-human-body.htm#0

Chapter 04

Diabetes : Natural Occurrence

In this chapter I shall attempt a study of diabetes that has occurred due to " natural " circumstance ; and that which has not been attributed to man - made Polychlorinated biphenyls .

What are the natural causes of diabetes . Let us first look at the definition by the Mayo Clinic :

" Causes of type 1 diabetes : The exact cause of type 1 diabetes is unknown. What is known is that your immune system — which normally fights harmful bacteria or viruses — attacks and destroys your insulin-producing cells in the pancreas. This leaves you with little or no insulin. Instead of being transported into your cells, sugar builds up in your bloodstream.

Type 1 is thought to be caused by a combination of genetic susceptibility and environmental factors, though exactly what many of those factors are is still unclear. " (1)

If we read into this , the cause is basically an attack by the immune system that destroys the insulin producing cells . However , this definition does not go into much detail on which parts of the immune system are involved , and exactly how the beta cells that produce insulin are attacked .

However , if we go back to the previous chapter , and look at the process as defined by the studies which were conducted by four different groups . Those scientists were :

1) Dencker L (1985). "The role of receptors in 2,3,7,8-tetrachlorodibenzo-p-dioxin (TCDD) toxicity

2) L. Poellinger. Mechanistic aspects – the dioxin (aryl hydrocarbon) receptor (2000). "Mechanistic aspects—the dioxin (aryl hydrocarbon) receptor.". Food Additives and Contaminants

3) J. Lindén, S. Lensu, J. Tuomisto, R. Pohjanvirta. (2010). "Dioxins, the aryl hydrocarbon receptor and the central regulation of energy balance. A review.". Frontiers in Neuroendocrinology

4) Okey, A. B. (2007). "An aryl hydrocarbon receptor odyssey to the shores of toxicology: The Deichmann lecture". International congress of toxicology-XI. Toxicological Sciences

And their conclusions were :

" The toxicity is mediated through the interaction with a specific intracellular protein, the aryl hydrocarbon (AH) receptor, a transcriptional enhancer, affecting a number of other regulatory proteins "
" TCDD binding to the AH receptor induces the cytochrome P450 1A class of enzymes which function to break down toxic compounds, e.g., carcinogenic polycyclic hydrocarbons such as benzo(a)pyrene "

So in these studies specify an exact enzyme (the cytochrome P450 1A class of enzymes) which acts to break down toxic chemicals .

To be sure , let us look at other definitions of what causes naturally occurring diabetes :

This from Web MD :

Cause :

" Type 1 diabetes develops because the body's immune system destroys beta cells in a part of the pancreas called the islet tissue. These beta cells produce insulin. So people with type 1 diabetes can't make their own insulin. " (2)

So Web MD is saying basically the same thing as the Mayo Clinic , only with a few more details . Let us look at another cause , as found on the internet :

From the University of Maryland Medical Center :

" In type 1 diabetes, the pancreas does not produce insulin. Insulin is a hormone that is involved in regulating how the body converts sugar (glucose) into energy. People with type 1 diabetes need to take daily insulin shots and carefully monitor their blood glucose levels." (3)

Let's look at one more . It's clear to me that these definitions are some what vague ; and the mainstream definition is more concerned with treating the symptoms rather than the cause .

One more , this from dLife :

" Type 1 Causes

What causes diabetes? Researchers have identified several genes associated with the development of type 1 diabetes. While the causes are complex and not completely understood, the prevailing belief about the etiology, or cause, of type 1 diabetes is that while someone may have a genetic predisposition for developing the disease, it takes an environmental trigger or series of triggers (e.g., virus, toxin, drug) to set off the autoimmune process that destroys insulin-producing beta cells of the pancreas. " (4) (Read more at source 4 ; listed below)

Ah ! Here at dLife we have a more detailed description , although none of their 4 risk factors goes into a great deal of detail on the genetic factors . One of their risk factors is listed as " Autoimmune diseases " ; but it simply says this :

" The presence of other autoimmune disorders, such as thyroid disease and celiac disease, raise the risk of type 1 diabetes. " (4)

So this does not really detail damage to the immune system ; nor does it detail exact cellular causes . What WAS mentioned (in the paragraph quoted above is that " ...... while someone may have a genetic predisposition for developing the disease, it takes an environmental trigger or series of triggers (e.g., virus, toxin, drug) to set off the autoimmune process that destroys insulin-producing beta cells of the pancreas . " (4)

Is PCB poisoning one of these environmental triggers ? Apparently that has been proven in multiple studies that it is . But what about before the introduction of PCBs in the 1880's , and of other toxins that were being poured into the environment around that same time period . Let's look at the very early history of diabetes .

The Early History of Diabetes

Here is a brief history of the disease as noted by the National Center for Endocrinolgy Information (the NCBI)

" Clinical features similar to diabetes mellitus were described 3000 years ago by the ancient Egyptians. The term "diabetes" was first coined by Araetus of Cappodocia (81-133AD). Later, the word mellitus (honey sweet) was added by Thomas Willis (Britain) in 1675 after rediscovering the sweetness of urine and blood of patients (first noticed by the ancient Indians). It was only in 1776 that Dobson (Britain) firstly confirmed the presence of excess sugar in urine and blood as a cause of their sweetness. In modern time, the history of diabetes coincided with the emergence of experimental medicine. An important milestone in the history of diabetes is the establishment of the role of the liver in glycogenesis, and the concept that diabetes is due to excess glucose production Claude Bernard (France) in 1857. The role of the pancreas in pathogenesis of diabetes was discovered by Mering and Minkowski (Austria) 1889. Later, this discovery constituted the basis of insulin isolation and clinical use by Banting and Best (Canada) in 1921. Trials to prepare an orally administrated hypoglycemic agent ended successfully by first marketing of tolbutamide and carbutamide in 1955. This report will also discuss the history of dietary management and acute and chronic complications of diabetes." (5)

So diabetes mellitus was noticed as far as 3000 years ago . Am I contending that Polychlorinated biphenyls are the cause of diabetes (both Type 1 and Type 2) ? No , I am not . What I am saying is that the original makers of Polychloruinated biphenyls (first discovered in 1865 as a coal tar byproduct , and later synthesized by German chemists in 1881 (6) . At any rate these first synthesizers accidentally (or purposely) stumbled on an excellent way of triggering the human immune system response to fighting toxins in the body . At this chemical (PCB) stayed in the body , and attached itself to certain cells long enough for there to be severe damage done . The question on my mind is :

' What in nature is the PCB molecule imitating ?

PCBs were first produced as " PCBs, originally termed "chlorinated diphenyls", were commercially produced as mixtures of isomers at different degrees of chlorination. (6)" It might be useful to study what types of chlorine - like substances are naturally occurring in nature .

First , many of the foods that we eat today are contaminated with PCBs (fish in particular) ; but I am looking for the very early history of toxins that caused the incidence and spread of diabetes .
What we are talking about in the causal stages of diabetes is the body's immune system attacking and destroying the beta cells . This is called ' auto immune disease ' because it is the body attacking and destroying it's own cells . (7) Normally, the immune system protects the body from infection by identifying and destroying bacteria, viruses, and other potentially harmful foreign substances. (&) But in a patient with auto immune disease , we are talking about the immune system destroying the body's own cells . So we are talking about diabetes as an auto immune disease .

Diabetes is a genetic disease , with genetic information being passed down from generation to generation . The method of this genetically passed information is :

" Certain gene variants that carry instructions for making proteins called human leukocyte antigens (HLAs) on white blood cells are linked to the risk of developing type 1 diabetes. The proteins produced by HLA genes help determine whether the immune system recognizes a cell as part of the body or as foreign material. Some combinations of HLA gene variants predict that a person will be at higher risk for type 1 diabetes, while other combinations are protective or have no effect on risk. " (7)

By the way , the NDIC also lists insulin as a substance , when introduced into the body in daily injections , as a substance the immune system recognizes as a hostile substance , and produces antibodies to fight it .

But , there must be a factor in the passing of incorrect information through the gene structure (the HLA gene) . PCBs have been found to be a destroyer of genetic information , so we are still back to the question of ' what were the early causes of diabetes ? '

The first cause of pre PCB toxins diabetes is : Viruses and infections : according to the NDIC :

" A virus cannot cause diabetes on its own, but people are sometimes diagnosed with type 1 diabetes during or after a viral infection, suggesting a link between the two. Also, the onset of type 1 diabetes occurs more frequently during the winter when viral infections are more common. Viruses possibly associated with type 1 diabetes include coxsackievirus B, cytomegalovirus, adenovirus, rubella, and mumps." (7)

So viruses can be a key cause in the development of diabetes , as they can cause errors to occur in the immune system , and within the genetic structure . They can also destroy the insulin producing beta cells by themselves .

Secondly ; dietary factors :

" For example, breastfed infants and infants receiving vitamin D supplements may have a reduced risk of developing type 1 diabetes, while early exposure to cow's milk and cereal proteins may increase risk. More research is needed to clarify how infant nutrition affects the risk for type 1 diabetes. " (7)

So cow's milk is a dietary risk factor for infants . (according to the NDIC)

Other foods , of course , lead to the development of diabetes . These include : highglycemic, trans fat– and saturated fat–rich, low-fi ber, phytonutrient-poor food choices . (8)

Thirdly ; there is genetics as a factor :

" Genes play a significant part in susceptibility to type 2 diabetes. Having certain genes or combinations of genes may increase or decrease a person's risk for developing the disease.

The role of genes is suggested by the high rate of type 2 diabetes in families and identical twins and wide variations in diabetes prevalence by ethnicity. Type 2 diabetes occurs more frequently in African Americans, Alaska Natives, American Indians, Hispanics/Latinos, and some Asian Americans, Native Hawaiians, and Pacific Islander Americans than it does in non-Hispanic whites. " (7)

What these ethnicity developed genes are still being studied , according to the NDIC :

" Recent studies have combined genetic data from large numbers of people, accelerating the pace of gene discovery. Though scientists have now identified many gene variants that increase susceptibility to type 2 diabetes, the majority have yet to be discovered. The known genes appear to affect insulin production rather than insulin resistance. Researchers are working to identify additional gene variants and to learn how they interact with one another and with environmental factors to cause diabetes." (7)

" Studies have shown that variants of the TCF7L2 gene increase susceptibility to type 2 diabetes. For people who inherit two copies of the variants, the risk of developing type 2 diabetes is about 80 percent higher than for those who do not carry the gene variant.1 However, even in those with the variant, diet and physical activity leading to weight loss help delay diabetes, according to the Diabetes Prevention Program (DPP), a major clinical trial involving people at high risk. " (7)

So , the TCF7L2 gene has been identified to increase susceptibility to type 2 diabetes . I have not focused so much on type 2 diabetes as I have on type 1 . Perhaps it is that I am a severe type 1 diabetic , and my focus has been mainly on that aspect of things . But , that being said , many of the dynamics of development of the disease are the same in both type i and type 2 diabetes .

As I mentioned earlier , I am not saying that PCB poisoning is solely responsible for the development of diabetes (both type 1 or type 2) . What I AM saying is that PCB poisoning has mimicked factors in nature that contributed to , and caused the development of the disease ; much as many things in our modern society has taken phenomenon already existing in nature , and either intensified it , or mimicked it with the use of machines . In the case of PCBs , and environmental poisons , it has intensified the development of diabetes to epidemic proportions .

(1) Mayo Clinic   :   Diabetes : Causes   :
http://www.mayoclinic.org/diseases-
conditions/diabetes/basics/causes/CON-20033091

(2) Web MD   :   Diabetes Health Center :   Cause   :
http://www.webmd.com/diabetes/guide/type-1-diabetes-cause

(3) University of Maryland Medical Center   :   Diabetes - type 1   :
http://umm.edu/health/medical/reports/articles/diabetes-type-1

(4) dLife   :   Type 1 Causes   :
http://www.dlife.com/diabetes/type-1/diabetes-causes

(5) National Center for Endocrinolgy Information   :   History of
diabetes mellitus.   :   http://www.ncbi.nlm.nih.gov/pubmed/11953758/

(6) Wikipedia   :   Polychlorinated biphenyl   :
http://en.wikipedia.org/wiki/Polychlorinated_biphenyl

(7) National Diabetes Information Clearinghouse (NDIC)   :   What
causes type 1 diabetes?   :
http://diabetes.niddk.nih.gov/dm/pubs/causes/

(8) Dr Hyman   :   Environmental Toxins , Obesity, and Diabetes: An
Emerging Risk Factor   :   http://drhyman.com/downloads/Diabetes-
and-Toxins.pdf

Chapter Five

General Electric and Trans Humanism

In the late 1920s ; General Electric continued on at breakneck speed .

New buildings had been built to house pattern shops for the huge Foundry building , a Gas Plant was constructed south of East Street .

In 1930 a young man from Russia , one K. K. Paluev , " .... played a leading role in the development of forced-oil cooling for large power transformers. This development was to be of great advantage in the design of large, high power transformers needed for various aspects of the war effort during World War II, as well as the large capacity transformers needed during the post-war period." (1)

This is , of course , worthwhile , and even noble goals - - - the winning of a war , the providing of electricity to cities after the conclusion of the war . But for the workers of that time , starting in 1932 when GE first started using Polychlorinated biphenyls to cool it's transformers , until 1977 , this enterprise was quite literally ' hell on earth ' . In fact , when I resided in Pittsfield in the 1970s , I heard this term ' hell on earth ' being used to describe various parts of the General Electric plant .

Was this all so bad ? After all , every job has it's ' risks ' . But - the question that I am asking myself , is : Was there a greater ' purpose ' in play here ?

And that ' greater purpose ' is a movement called ' Transhumanism ' .

What exactly is Transhumanism ?

According to the Wikipedia website it is : " An international cultural and intellectual movement with an eventual goal of fundamentally transforming the human condition by developing and making widely available technologies to greatly enhance human intellectual, physical, and psychological capacities . " (2)

This term ' transhumanist ' was first espoused by the author , teacher , philospher and futurist Fereidoun M. Esfandiary (3) ; who taught "new concepts of the human" at The New School in the 1960s, when he began to identify people who adopt technologies, lifestyles and worldviews "transitional" to posthumanity as "transhuman ." (2)

Exactly how far back does the transhumanist movement go ? Wikipedia : " According to Nick Bostrom,[1] transcendentalist impulses have been expressed at least as far back as in the quest for mmortality in the Epic of Gilgamesh, as well as in historical quests for the Fountain of Youth, the Elixir of Life, and other efforts to stave off aging and death." (2) So it is man's attempt to stave off aging and death .

So how does the General Electric fit into this ' transhumanistic ' scenario ? To find this out , we must delve a bit deeper into the tenets of the philosophy . There have been ten tenets , or ' top ten technologies ' identified with this movement .

These according to the Lifeboat Foundation (Safeguarding Humanity) are :

" 10) Cryonics

Cryonics is the high-fidelity preservation of the human body, and particularly the brain, after what we would call death, in anticipation of possible future revival. Cryonics is an important transhumanist technology not only because it is already available today, but because the technology is relatively mature — we can reliably stop cells from decaying.

9) Virtual Reality

Sometime in the 2020s, reality simulations will become so high-resolution and immersive that they'll start to get indistinguishable from the real thing.
Simulations will become the preferred environments for work and play. Pretty soon the main obstacle to truly immersive VR will not be the visuals but the haptics — our sense of touch. To fool our senses into believing haptic technologies are conveying the real thing, the "frame rate" needs to be significantly higher than for visual technologies, a few hundred updates per second rather than a few

dozen — which is why development could take another decade or two. But many millions of dollars are currently going into efforts to develop advanced VR.

8) Gene Therapy/RNA Interference

Gene therapy replaces bad genes with good genes, and RNA interference can selectively knock out gene expression. Together, they give us an unprecedented ability to manipulate our own genetic code.

7) Space Colonization

Space colonies will become necessary to house the many billions of individuals that will be born in the future as our population continues to expand at a lazy exponential. In his book, The Millennial Project, Marshall T. Savage estimates that the Asteroid Belt could hold 7,500 trillion people, if thoroughly reshaped into O'Neill colonies. The O'Neill cylinder (also called an O'Neill colony) is a space settlement design proposed by American physicist Gerard K. O'Neill in his 1976 book The High Frontier (4)

6) Cybernetics

Cyborgs already walk among us, and they look just like normal people.

This trend will continue in the future. Many cyborg upgrades which will become available in the 20s and 30s, such as hearing and vision enhancement, metabolic enhancement, artificial bones, muscles, and organs, and even brain-computer interfaces will be invisible to the casual observer, implanted beneath the skin.

5) Autonomous Self-Replicating Robotics

Why do manual labor when the robots can do it for you? Self-replication might be considered the Holy Grail of robotics. A landmark NASA study, "Advanced Automation for Space Missions", found that robotic self-replication is just a matter of engineering, and that no fundamental theoretical breakthroughs are needed.

4) Molecular Manufacturing

If self-replication is the Holy Grail of robotics, then molecular nanotechnology (MNT) is the Holy Grail of manufacturing. Molecular nanotechnology would use massive arrays of nanometer-scale actuators (produced initially through self-replication) to manufacture macroscale products with atomic precision. This concept is known as the nanofactory.

3) Megascale Engineering

Most people are familiar with megascale engineering because it is seen throughout fiction — the Death Star, for instance. Typically, megascale engineering refers to building structures at least 1,000 km in length in one dimension, such as a space elevator, Globus Cassus (an art project and book[1] by Swiss architect and artist Christian Waldvogel presenting a conceptual transformation of Planet Earth into a much bigger, hollow, artificial world with an ecosphere on its inner surface (5) , or Dyson sphere (a hypothetical megastructure that completely encompasses a star and hence captures most or all of its power output. (6)

2) Mind Uploading

Mind uploading, sometimes referred to as nonbiological intelligence, centers around the controversial proposition that cognitive processing can be implemented on substrates other than our current neurons. Considering decades of successful results in neurophysiology, and the recent construction of the world's first brain prosthesis — an artificial copy of the hippocampus — this seems very likely.

1) Artificial General Intelligence (AGI)

As argued in the previous section, functionalism seems likely. If so, then strong AI is possible. Thinking, feeling, imagining, creating, communicating, thoughtful synthetic intelligences with conscious experiences. Whether serial computing is sufficient, or parallel computing is necessary, both are within technological reach, and present-day computing speeds are fast approaching the computing power of the human brain. "

Note : I have provided just short quotes of the subjects above . If one wishes to read further on any of these subjects visit the Lifebioat website at : http://lifeboat.com/ex/transhumanist.technologies

Some of these potential innovations are good ideas - - - even grand in their scale . The questions that we must ask here are : ' What has the human cost been , leading up to these ' grand experiments ' ; and secondly , what has the role of the General Electric been in the field of transhumaism .

I must say that in my own experience , being a Type 1 diabetic and an amputee , who has had multiple issues with diabetes , and whose diabetes has been described as ' very severe ' , I began to ask questions ; not so much ' how ' (I have covered that part in previous chapters) , but ' why ' . I shall now attempt to look at the ' why ' .

So , how many of these ' top ten ' transhumanist technologies is GE actually involved in ?

For one , General Electric is heavily involved in Virtual Reality (number 9 of the top ten) . The GE has built 3D simulators , and has included the VR experience in it's new $ 500 million dollar undersea research project . (8) I must say that General Electric , nowadays , is looked at as being ' old technology ' ; and thus they are hurrying to catch up with many of the newer tech companies .

Let us now look at number 8 on the top ten list , Gene Therapy/RNA Interference . How is GE involved here ? General Electric has a subsidiary of it's company which is called ' GE Healthcare Life Sciences ' . RNA is a gene which communicates with hundreds of other genes simultaneously , and they " ... induce subtle but reproducible shifts in gene expression. " (9) But , isn't the use of PCB's , and the resultant toxicology studies made on humans that were infected by this toxic substance a form of gene therapy and RNA interference ? After all , the genes that cause diabetes are not receiving the correct signals from the immune system when infected with PCBs and the various PCB by - products .
The lack of a correct signal from the immune system is what causes the cytochrome P450 1A class of enzymes which function to break

down toxic compounds , to attack the beta cells (which produce insulin) , falsely identifying them as a ' foreign substance ' . Once again I must refer to the fact that GE is thought of as ' old technology ' , and their studies of toxic effects of certain chemicals on genes , and communication between genes must be looked at as being in a very old era of gene research (the 1930's , 1940's and 1950's) . It was during this same era (the 1930's and 1940's) that Nazi Germany was doing massive toxicology experiments and research on concentration camp inmates / I just thought that I would point that out . Our methods of study on interruption of signals between genes has come a long way since that time .

   Well , let us continue to move down the list . Number 7 on the list is Space Colonization . According to the Space Settlement Institute , General Electric is one of the publicly traded companies that is heavily invested in space settlement . (10) Number 6 on the list , Cybernetics , is something that GE is heavily invested in , dating back to 1966 , when GE was commissioned to build a cybernetic walking machine for the U.S. Army . It was built and delivered in 1970 . (11) Number 5 on the list , Autonomous Self-Replicating Robotics , is a newer field that GE , as a parent company , does not have a lot of experience with , although they have quite a bit of experience in the field of robotics , including the DP 6xx Robot System (12) , and a robotic enabled intelligence system (13) .

   Moving on to number 4 on the list , Molecular Manufacturing , or nanotechnology , is a field that GE is nvested in . In 2014 the State of New York invested $ 135 million , along with GE to " ... spur high-tech manufacturing of miniature electronics . " (14) Another $ 365 million will be invested by GE and other companies , bringing the total invested into this field to $ 500 million . Continuing on to number 3 , in the case of Megascale Engineering , it is difficult to find the GE involved in any of these large scale projects , remembering that building structures at least 1,000 km in length in one dimension is very rare . However , one can bet that there would be some interest in the Dyson Sphere , which involves the construction of a large sphere that would envelop and capture the energy of a local sun (within this galaxy) . But for now I must say that the General Electric is not highly involved in the area of Megascale Engineering . After all , this is a highly theoretical field at this particular time .

On to number 2 on the list , Mind Uploading , or ' transtopia ' GE is highly involved in the area of ' minds and machines ' , and in October 2014 hosted a conference on minds and machines , with the current CEO of GE , Jeff Immelt , hosting leaders from business and technology to discuss the subject . This lines up fairly accurately with Adam Kadmon's definition from the Transtopia website : " Uploading is the transfer of the brain's mindpattern onto a different substrate (such as an advanced computer) which better facilitates said entity's ends. Uploading is a central concept in our vision of technological ascension . " (16) . This is an important factor in the transhumanist philosophy . The machine must be able to replicate , and actually out-perform the human mind .

On to number 1 on the list , Artificial General Intelligence (AGI) . Here GE is heavily involved in this type of research . This from AI (Artificial Intelligence) Magazine : " General Electric is engaged in a broad range of research and development activities in artificial intelligence, with the dual objectives of improving the productivity of its internal operations and of enhancing future products and services in its aerospace, industrial, aircraft engine, commercial, and service sectors. Many of the applications projected for AI within GE will require significant advances in the state of the art in advanced inference, formal logic, and architectures for real-time systems. " (17) This is very important for the intelligent machine production methods currently employed by GE . Machines will be the ultimate production workers .

There you have a quick rundown of GE's involvement in transhumanist technologies . Now , let me say that some of these projects are very interesting , and may prove to be of some benefit to mankind . But , if they are handled in the same haphazard manner that GE used in it's transformer production , and with the same disregard for the natural environment that GE showed in it'd misuse of the chemical PCB , mankind is in quite a bit of trouble . Maybe what GE and their cohorts are building is a world where only machines are able to survive in the environment that GE creates .

Yes , General Electric was run out of , or forced to close , I might say , to use a gentler term ; in most locations in the northeast U.S. , due to the harsh environmental regulations and fines that were levied upon

them , and they were forced to move many of their operations overseas
.

But , has GE learned it's lessons from it's harsh treatment of the environment ? Let us take a look .

(1) IEEE Global History Network : Archives:Transformers at Pittsfield, part 1 : Biographies : http://www.ieeeghn.org/wiki/index.php/Archives:Transformers_at_Pittsfield%2C_part_1

(2) Bostrom, Nick (2005). "A history of transhumanist thought" Journal of Evolution and Technology. Retrieved 2006-02-21 : http://www.nickbostrom.com/papers/history.pdf

(3) Wikipedia : FM-2030 : http://en.wikipedia.org/wiki/FM-2030

(4) Wikipedia : O'Neill cylinder : http://en.wikipedia.org/wiki/O'Neill_cylinder#Island_Three

(5) Wikipedia : Globus Cassus : http://en.wikipedia.org/wiki/Globus_Cassus

(6) Wikipedia : Dyson Sphere : http://en.wikipedia.org/wiki/Dyson_sphere

(7) Lifeboat : Special Report : Top Ten Transhumanist Technologies : http://lifeboat.com/ex/transhumanist.technologies

(8) Road to VR : GE Marks $500M Subsea Research Project Launch With VR Experience from Studio Kite & Lightning : http://www.roadtovr.com/tag/general-electric-vr-experience/

(9) GE Life Sciences : RNA Interference : http://www.gelifesciences.com/webapp/wcs/stores/servlet/catalog/en/GELifeSciences-th/applications/rna-interference/

(10) The Space Settlement Institute : Publicly Traded Space Companies : http://www.space-settlement-institute.org/space-companies.html

(11) U.S. Army Transportation Museum : Cybernetic Walking Machine :
http://www.transchool.lee.army.mil/museum/transportation%20museum/cybernetic.htm

(12) AZO Robotics : DP 6xx Robot System from General Electric Company : http://www.azorobotics.com/equipment-details.aspx?EquipID=435

(13) 4-Traders : General Electric Company : GE to Develop Robotic-Enabled Intelligent System Which Could Save Patients Lives and Hospitals Millions : http://www.4-traders.com/GENERAL-ELECTRIC-COMPANY-4823/news/
       General-Electric-Company--GE-to-Develop-Robotic-Enabled-Intelligent-System-Which-Could-Save-Patient-15982883/

(14) Yahoo Finance : NY invests in nanotech with General Electric : http://finance.yahoo.com/news/ny-invests-nanotech-general-electric-194557619.html;_ylt=A0SO81Xz1dtUCMUA2uZXNyoA;_ylu=X3oDMTEzZjIxbTQ5BGNvbG8DZ3ExBHBvcwMxBHZ0aWQDVklQNTYyYXzEEc2VjA3Ny

(15) GE Minds and Machines : Minds + Machines 2014 : https://www.gemindsandmachines.com/

(16) Transtopia : Transhumanism Evolved : Mind Uploading - An Introduction : http://transtopia.net/uploading.html

(17) AI Magazine : Artificial Intelligence Research at General Electric :
http://www.aaai.org/ojs/index.php/aimagazine/article/viewArticle/502

Chapter Six

Continued Environmental Destruction

In the last chapter I detailed some of the tenets that the General Electric corporation follows that would qualify it as a transhumanistic company . I have mentioned this , perhaps , for my own reasons , in that when I pursued answers to the question : " Why ? " , rather than the " How " (which I also attempted to pose some answers for in the previous chapters) ; at any rate , when I asked the question " Why ? " the terrible experimentation that caused so much suffering here in the United States , I was always given the answer : ' That it was for the greater good ; it was for the coming of ' the machine - man ' , the combination of man and machine , that would rule the world with no worries .

However , we (or GE , at any rate) must be far short of that goal yet ; as they have continued their environmental destruction , and disregard for human health overseas . This after being run out of the United States by restrictive environmental protection laws and huge fines ; which made the open, or nonexistent environmental protection laws in other countries , along with the lower wages that they would be able to pay the workers , a very tempting and profitable proposition .

So , I , and we , are beginning to see that GE certainly has **NOT** learned it's lesson . The first example that we have is allowing Chinese workers to build it's compact fluorescent light bulbs (CFLs) , which contain mercury , in unsafe conditions . Barrels containing mercury were left open in the plant , they were not labeled properly , and the workers had little or no training on how to handle them properly . Thus , many workers at the plant in Xiamen, Fujian Province , China came down with some form of mercury poisoning - - numbness in the legs , swollen fingers , numbness all over their bodies , some reported . (1) And on top of this there were multiple violations of China's labor laws at the plant . These include :

" • Requiring work hours that are longer than the permitted maximum average in 2007 of 203.4 hours a month;

• Providing no pay stubs, so workers can't tell if they are being correctly paid;

• Not paying overtime for work in excess of 8 hours a day or on the sixth day of work each week, which under Chinese law is to be a day of rest; and

• Mandating that workers who quit without permission forfeit a month's wages.

These violations of Chinese labor law also infringe GE's own corporate policies, which call for the company to obey local laws and expect suppliers to "comply with laws and regulations governing minimum wages, hours of service and overtime wages for employees." Most of them also contravene the Electronic Industry Code of Conduct of leading electronics companies. " (1)

Moving to another sector that GE is particularly fond of , that of ' Power Africa ' (a project that GE heavily backs that will " "develop an appropriate mix of power solutions" to provide electricity, fight poverty and "drive economic growth." (2) This sounds good , but .... " The real concern here is that US taxpayers will wind up supporting African energy development that caters to corporate industrial zones and natural resource exporters, leaving the majority of Africans in rural and neglected urban areas still without access to power and exposed to dangerous pollution." (2) To continue on the Power Africa problems :

" An OPIC proposal to finance the Azura Edo gas plant in Nigeria is a recent case in point. A review of project documents and site visits by Environmental Rights Action Nigeria found that the plant will not provide any new energy access, even to villages immediately adjacent to the project, nor will families displaced by the project receive adequate resettlement compensation. Project developers did not consider any renewable energy options. Instead, the plant will use open-cycle gas turbines supplied by GE—a technology more polluting and less energy efficient than closed-cycle turbines. Yet this project is considered part of Power Africa." (2)

This article on Power Africa is a very good read , as it has a section on the ' Global Climate Justice Movement .' (see this at link 2 below) . The people residing in the poorer neighborhoods " understand firsthand the effects of energy pollution and climate chaos . " (2)

Another area where GE still operates in the United States is the controversial area of fracking for shale oil . GE is investing billions , according to an Ohio.com report . (3) For those unfamiliar with fracking , it is " .... the process of pumping a combination of sand, water, and chemicals into the ground. This increases pressure underground and causes fracturing to occur, the purpose of which is to release hidden stores of oil and natural gas from within the rock layer that is fractured. " (4) However , all is not happiness in fracking land , as the same report states : " it has recently come to light that fracking can pose serious dangers and health risks to residents living in surrounding areas." (4) The most serious problem is the contamination of water by the chemicals used in the fracking process . This , however , is of no concern to GE , as they have a great deal of experience in chemical pollution .

Another area to consider is the area of depleted uranium . Depleted uranium (DU) was not directly manufactured by the General Electric Corporation . Who did manufacture it was the General Dynamics Company . (6) What General Electric DID do , however , is manufacture and supply the delivery system for the rounds of DU , the GAU-8/A cannon . This cannon was mounted on the A-10 warthog fighter plane , and delivered 320 to 750 tons of DU in Iraq 1991 , 10 to 200 tons in the Balkans 1999 . and 500 to 600 tons in Afghanistan . (7) What is depleted uranium exactly ? It is : " is uranium with a lower content of the fissile isotope U-235 than natural uranium.[2] (Natural uranium is about 0.72% U-235 . " (8) It is used in artillery shells at times (GAU-8/A cannon) because of it's high penetration index (due to it's radioactivity .) However , once DU has been fired it emits U - 238 . " Dr. Rosalie Bertell presented a concise explanation of the potential dangers of exposure to depleted uranium (DU). Dr. Bertell stated : "Uranium oxide and its aerosol form are insoluble in water. The aerosol resists gravity, and is able to travel tens of kilometres in air. Once on the ground, it can be resuspended when the sand is disturbed by motion or wind. Once breathed in, the very small particles of uranium oxide, those which are 2.5 microns (one micron = one millionth of a meter) or less in diameter, could reside in the lungs for years, slowly passing through the lung tissue into the blood." (9) " The Department of Defense continues to deny health risks associated with the use of DU's, yet it's own actions belie their claims. May 15th of 2003 found Scott Peterson of the Christian Science Monitor reporting that in Iraq, "Six American vehicles struck with DU "friendly fire" in 1991 were deemed to be too contaminated to take home, and were buried in Saudi Arabia. Of 16 more brought back to a purpose-built facility in South Carolina, six had to be buried in a low-level radioactive waste dump." (9) And how long does DU stay in the environment ? Depleted Uranium is 40 % less radioactive than uranium , and uranium stays
in the environment for a short time , but it decays to form U-234 , U-235 and U-238 . What are the health effects of these ? " Intakes of uranium exceeding EPA standards can lead to increased cancer risk, liver damage, or both. Long term chronic intakes of uranium isotopes in food, water, or air can lead to internal irradiation and/or chemical toxicity." (10) So just another study in toxicity for GE . And , of course , the people most affected by this depleted uranium are the poor people of Iraq and Afghanistan , where birth defects have climbed enormously . (11) And use of depleted uranium is a violation of international law . (12) Legality and General Electric ? They , and their partners such as General Dynamics and Monsanto , have a long history of flaunting the law .

And depleted uranium is very difficult to clean up . Let us take a look at some of the toxins that General Electric is responsible for that possibly CAN be cleaned up .

To be fair , GE has attempted to (been forced to) involve itself in the clean up of some of it's more toxic sites ; including the Hudson River PCB spill , the Pittsfield area , the Schenectady , NY area , the Anniston , Alabama site (which is reportedly the worst of the bunch , read more at the linked article below .) Of course now the entire world is being forced to look at China's pollution , because this type of damage to the water , food chain and ozone layer could have worldwide effects , and now GE has climbed aboard the bandwagon to ' assist ' in the cleanup .

But how effective are these General Electric cleanups ?

I know , from personal experience , of having been in the Pittsfield , Massachusetts area in 1997 and 1998 (when the controversy about PCB clean up , and the pollution of residential properties was just breaking) that this clean- up was not very effective at all . For one , PCBs kept cropping up in unexpected areas , such as " ... a site on Newell Street where 250 barrels have been found, 21 of which were still intact and contained oil that was as much as 78 percent PCBs . " (13) (Note : the lot adjacent to this land parcel on Newell Street is known to be very heavily polluted , GE having pumped 36,000 gallons of PCB oil out of the ground since 1999 (13) . This site , though , is the project of a plan to blacktop the lot , and place filler soil and sod on top of that .

For another , there are STILL problems with the PCB clean up of Silver Lake (across the street from the GE plant from the 1920s though the exit of General Electric in the early 1990's . In this internet article , dated in April 2012 , Silver Lake (which we called ' black lake ' in the 1970s , due to the color that the pollution had turned it) was being argued about as to which contractor would be named to do the job . And this being 33 years after PCBs were outlawed (1979) , and 14 years after the GE was ordered to begin the clean up of the Pittsfield site . Source (14)

But , the worst of the clean-up ' incidents ' had to be the creation of a toxic waste dump next to an elementary school , Allendale Elementary , in the Pittsfield suburb of Allendale . The story of how PCB sludge originally got next to , and onto the school grounds is terrible enough - - - " When you have a dangerous chemical such as PCB, how do you dispose of it? If you're General Electric, you bury it on site and around the city. GE also dumped PCBs and other chemicals into Unkamet Brook and Silver Lake in Pittsfield, and when they still had more to get rid of, they offered it to their neighbors, their own workers, the city, and even to their community schools. According to the US Environmental Protection Agency, from the 1940s through the 1970s, GE gave away thousands of tons of fill from its facility to Pittsfield-area homeowners and contractors. GE's PCBs have turned up in the backyards of former workers, in city parks and playgrounds, and even in city schoolyards. In 1997, GE was ordered to remove 3,800 cubic yards of contaminated soil from a city playground, Dorothy Amos Park. The Park has been remediated, but the river bank and river next to the park still have high levels of PCBs." (15) But this is not the whole story . When GE was ordered to finally start the clean up of the Housatonic River , they took contaminated sludge that they had dredged from the river , and dumped it next to the elementary school once again . Then , to make matters worse , the then mayor of Pittsfield , one Gerald (Jerry) Doyle , signed a pledge to cap ... rather than clean the toxic PCB site next to the Allendale Elementary . The problem with capping is that the caps only last from one day to a maximum of 25 years . (this according to blogger Jonathon Melle : Source (16) Once those caps for the toxic waste goes bad , well , then Pittsfield has problems , not only with the children next to the dump site , but also with the waste getting into land and water .

Of course , I have been only concentrating my report mainly on the PCB problem in Pittsfield , and mostly as it relates to Type 1 diabetes . General Electric is also quite guilty of being a major contributor to asbestos pollution in the area , and , of course , to cancer . Blogger Jonathon Melle's mother died of cancer that he believes was caused by PCBs .

So what can we say that would distinguish GE from many of the other large corporate polluters , not only in the U.S. , but in the world ? According to the website ' Essential Information ' (Essential information is a watchdog website of General Electric's misdeeds) : " IV. Conclusion : The Corporate Crime Reporter recently released a list entitled: The Top 100 Corporate Criminals of the Decade. Using a narrow definition that only included corporations that pled guilty or no contest to crimes and that have been criminally fined, the survey found that General Electric was number 34 on the list. "
" What distinguishes General Electric is not merely the number of crimes committed -- or the dollar amount of the crimes -- but a consistent pattern of violating criminal and civil laws over many years. Even worse, General Electric has been a leader in using political influence to attempt to overturn environmental and defense contracting laws that GE persistently violates. " (17)

And when General Electric is not openly violating and flaunting environmental law ; when a particular country , such as the United States , finally cracks down and attempts to force GE to clean up it's act , General Electric is always looking for another country or region where the environmental regulations are not so rigid , and where the workforce is more plient to GE's way of doing things .

(1) Daily Kos : General Electric: Killing Chinese Workers For A Cleaner Environment : http://www.dailykos.com/story/2008/04/01/488145/-General-Electric-Killing-Chinese-Workers-For-A-Cleaner-Environment#

(2) The Nation : What's Wrong With the Electrify Africa Act : http://www.thenation.com/blog/179740/whats-wrong-electrify-africa-act

(3) Ohio.com : General Electric spending billions on fracking research : http://www.ohio.com/blogs/drilling/ohio-utica-shale-1.291290/general-electric-spending-billions-on-fracking-research-1.401605?cache=18961415304345%252525252525252f

(4) Amaro Law Firm : Dangers of Fracking : http://www.oilgaslawsuits.com/dangers-of-fracking/

(5) Catastrophe Map : Anniston AL Contamination Monsanto : http://catastrophemap.org/toxic-apocalypse-anniston-alabama.html

(6) Ban Depleted Uranium : International Coalition to Ban Uranium Weapons : http://www.bandepleteduranium.org/en/a/72.html

(7) Cursor dot org : Uranium Wars: The Pentagon Steps Up its Use of Radioactive Munitions : http://cursor.org/stories/uranium.htm

(8) Wikipedia : Depleted uranium : http://en.wikipedia.org/wiki/Depleted_uranium

(9) Disabled - World.com : Dangers and Health Effects of Depleted Uranium : http://www.disabled-world.com/health/uranium.php

(10) EPA.gov : Radiation protection : Uranium : Health Effects of Uranium : http://www.epa.gov/rpdweb00/radionuclides/uranium.html

(11) RT.com : Depleted uranium used by US forces blamed for birth defects and cancer in Iraq : http://rt.com/news/iraq-depleted-uranium-health-394/

(12) Counter Currents.org : Depleted Uranium And International Law : http://www.countercurrents.org/du-shah231004.htm

(13) Berkshire Blog : Topic: PCB's and other chemical pollution :
http://berkshireeagle.blogspot.com/2005/03/topic-pcbs-and-other-chemical.html

(14) The Berkshire Eagle : Pittsfield PCB cleanup at Silver Lake postponed
: 04/06/2012 : http://www.berkshireeagle.com/local/ci_20337862
        /pittsfield-pcb-cleanup-at-silver-lake-postponed?source=rss

(15) Berkshire Environmental Action Team : General Electric and PCBs :
http://www.thebeatnews.org/BeatTeam/ge-pcbs/

(16) Jonathon Melle on Politics :
GE...Pittsfield...PCBs...CANCER...corruption... :
http://jonathanmelleonpolitics.blogspot.com/2009/05/gepittsfieldpcbscancercorrup
tion.html

(17) Clean Up GE .org : GE Misdeeds : Conclusions :
http://cleanupge.org/gemisdeeds.html

Chapter Seven

The Future of Diabetes Care

So what does the future hold for diabetes care ? (both Type 1 and Type 2) This disease , of which the Polychlorinated Biphenyl dumping in various parts of the world , did so much to propagate ?

I do believe that a large amount of evidence now points to the fact that GE knew , as early as the 1930s , that PCBs caused a cocktail of diseases . But , let us focus our attention , for the time being , on Type 1 diabetes .

Looking first at the transplantation of beta cells . This seems to be the ' state of the art ' procedure that is now being used to treat type 1 diabetes . At present the way that beta cells are transplanted (in order to give maximum benefits in maintaining long-term metabolic control in type 1 diabetic recipients . (1) At this point in time , a mass , or graft of at least " 2 million beta-cells per kg of body weight were needed to achieve signs of functioning grafts . " (1)

This sounds all well and good , yet there remain problems .

First , beta cells are found in the pancreas ," an organ about the size of your fist, that sits behind the stomach. " (2) Only about 2 % of the pancreas is made up of beta cells . The problems arise when beta cells are removed from the pancreas . It is very easy to injure , or destroy them . (2) Even if they are removed successfully " finding the appropriate material to surround the intact beta cells or islets to keep the immune system from destroying these cells as soon as they are introduced into the body is difficult." (2) A third problem arises when " .... stem cells are gradually lost and need to be replaced.Keeping the encapsulated cells alive as long as possible reduces the frequency of surgical replacement." (2)

So we are faced with a triad of problems when dealing with beta cell replacement ; which are the reasons why beta cell transplantation is not more prevalent .

But , let us recall our look at the destroyer of the beta cell , the cytochrome P450 1A class of enzymes . (3) Is there something to be gained from looking at this class of enzyme ? An enzyme that is part of the immune system that destroys intruders in the body ; which we learned that is what the bodies' immune system identifies polychlorinated biphenyl as .

The questions are : what are the similarities between pcb poisoning , and a ' natural ' occurring phenomenon , such as a virus infection ; and then , what steps can be taken to shield the beta cell from the P450 1A class of enzymes (in terms of encapsulization) .

A further look is needed to define this more closely . it is the binding of a particular class of toxin , known as TCDD (or Tetrachlorodibenzodioxin) , which is
" ..... usually formed as a side product in organic synthesis and burning of organic materials. " (4) It is this TCDD that binds to an AH receptor (an AH receptor is " a protein that in humans is encoded by the AHR gene. The aryl hydrocarbon receptor is a ligand-activated transcription factor involved in the regulation of biological responses to planar aromatic hydrocarbons. This receptor has been shown to regulate xenobiotic-metabolizing enzymes such as cytochrome P450. " (5) It is the interaction between TCDD , and the AH receptor that causes the P450 1A class of enzymes to go into action .

Is there a way to block the AH receptor ; or to encode the AHR gene not to react ? Would this be dangerous to the body , and to the immune system ?

I think the thing to do here would be to take a look at how genes , the genetic structure and cells communicate with one another .

Simple cellular communication takes place via a hormone that is released ; a chemical messenger , if you will . (6) The target cell receives the message through proteins indented into it's membrane called receptors . Thus , we see that the AH receptor is one of these proteins that facilitates communication between cells . Is there damage that was done to this AH receptor (through viruses , or toxic poisoning) that causes it to send incorrect messages that causes the immune system to attack the beta cells that produce insulin ?

Further : " This process of cellular communication is known as signal transduction. The end result of signal transduction, and a key step in cell decision making, is the switching on, or off, of protein production – more commonly called gene expression. Another class of membrane proteins that aid in cellular communication are channel proteins. These proteins act as molecular gates that allow the passage of small molecules and ions ; for example, glucose and sodium, across the membrane in response to a stimulus, such as an electrical current in the case of ions or insulin signaling in the case of glucose." (6)

Let us look at the phenomenon of ' gene expression ' a bit closer . This , according to Wikipedia : " Several steps in the gene expression process may be modulated, including the transcription, RNA splicing, translation, and post-translational modification of a protein. Gene regulation gives the cell control over structure and function, and is the basis for cellular differentiation, morphogenesis and the versatility and adaptability of any organism. Gene regulation may also serve as a substrate for evolutionary change, since control of the timing, location, and amount of gene expression can have a profound effect on the functions (actions) of the gene in a cell or in a multicellular organism." (7)

Continuing : " The genetic code stored in DNA is "interpreted" by gene expression, and the properties of the expression give rise to the organism's phenotype. Such phenotypes are often expressed by the synthesis of proteins that control the organism's shape, or that act as enzymes catalysing specific metabolic pathways characterising the organism." (7)

So what we have here is a problem of communication between the genetic code , and the way that the gene expression interprets that information , and then sends that information on to the cells , for function within the phenotype of the organism . (Note : ' phenotype ' meaning : " A phenotype (from Greek phainein, meaning "to show", and typos, meaning "type") is the composite of an organism's observable characteristics or trait ... " (8) Apparently , we need to break this ' communication ' down a bit further .

Researchers at the EMBL (European Molecular Biology Laboratory) have studied a gene called ' Fgf8 ' , which is responsible for communicating with cells , and developing a healthy , and well proportioned animal . The researchers " have elucidated how Fgf8 in mammal embryos is, itself, controlled by a series of multiple, interdependent regulatory elements . " (9)

Further : " Fgf8 is controlled by a large number of regulatory elements that are clustered in the same large region of the genome and are interspersed with other, unrelated genes. Both the sequences and the intricate genomic arrangement of these elements have remained very stable throughout evolution, thus proving their importance. By selectively changing the relative positioning of the regulatory elements, the researchers were able to modify their combined impact on Fgf8, and therefore drastically affect the embryo." (9) Note : A couple of definitions : 1) ' genome' is defined as " In modern molecular biology and genetics, the genome is the genetic material of an organism. It is encoded either in DNA or, for RNA viruses, in RNA. The genome includes both the genes and the non-coding sequences of the DNA/RNA." (10) And ' regulatory elements ' as " are regions of non-coding DNA which regulate the transcription of nearby genes." (11)

It is the regulatory elements , or regions that I wish to take a look at . The researchers at EMBL were able to change the positioning of the regulatory elements , the researchers were able to modify their effect on the gene communicator Fgf8 , and thus drastically affect an embryo . Thus , it is nothing more than a re-positioning of these regulatory elements .

Is this all that it is going to take to affect the broken signals that are being sent by AH receptors to the immune system that destroy the beta cells that produce insulin . It may not be encapsulating , but repositioning the regulatory elements to send proper signals to the beta cell . This is one possibility .

To say something further on the genetic communications subject , man has fallen from his lofty place of possibly having 12 strands of DNA , to having , at present , 2 strands . (12) We are now taking a second look at DNA within the body that has previously been dismissed as ' junk ' , and wondering ' how many strands can a human being actually possess ? ' Just a few years ago a two year old child was found with 3 DNA strands . (12) Does the future of genetic communications lie in positioning the regulatory elements in these newly found DNA strands ? Is this a question of human development ? Will ' future man ' be a Super Man , with geneticists able to regulate diseases out of the body ? Or , to take a slightly different tact , did mankind once possess 12 strands of DNA , and through some type of ' accident ' , disease , manipulation of some type (?) now only possess two .

Dr . Bernard Fox , one of the Cambridge researchers involved in this study , claims that other DNA helixes are being formed in our current double helix DNA . (12) He says that we are ' evolving ' , and that eventually we will develop the 12 strand DNA helix formation . Research in this area is still very early , but Russian researchers are claiming that communication between our genes takes place much the same as our actual spoken language (12) . So , thus , we are breaking down to the very basics how genes communicate their information to one another . According to this Russian research project " They found that the alkalines of our DNA follow a regular grammar and do have set rules just like our languages. So human languages did not appear coincidentally but are a reflection of our inherent DNA. " (12)

This from the Russian physicist and molecular biologist Pjotr Garjajev and his colleagues : " Living chromosomes function just like solitonic/holographic computers using the endogenous DNA laser radiation." This means that they managed for example to modulate certain frequency patterns onto a laser ray and with it influenced the DNA frequency and thus the genetic information itself. Since the basic structure of DNA-alkaline pairs and of language (as explained earlier) are of the same structure, no DNA decoding is necessary. " (12) So now we see that researchers are beginning to be able to use lasers and the knowledge of holographic computers to decipher , and possibly communicate with genes within the DNA structure . Will the scientist of the future be able to tell , let us say , the broken signals that are now being sent by the AH Receptors to the immune system , which are misinterpreted by P450 1A class of enzymes to attack and destroy the beta cells which produce insulin ?

The question on my mind would be : what caused the damage to the genetic communication system between the genetic system and the immune system in the first place ? I have touched on this a bit previously ; it is caused by viruses , and outside toxins such as Polychlorinated Biphenyls . Thus , we must ask ourselves , how far away are we from developing a new communication system between genes in the cellular and DNA system ?

Rather than deviate off , at this point , and look at all of the genetic communication research that is taking place today , I wish to focus on the Garjajev study , as this seems to offer a unique perspective . Dr. Garjajev and his team have presupposed the existence of an ' invisible web ' of information that our DNA is in contact with .... a web that such people as psychics , clairevoyants , people who's DNA is ' turned on ' , to exchange ideas over vast distances mentally . (13) " Russian researcher Dr.Vladimir Poponin put DNA in a tube and beamed a laser through it. When the DNA was removed, the laser light continued spiralling on its own, like it would through a crystal! This effect is called 'Phantom DNA Effect'. " (13)

" It is surmised that energy from outside of space and time still flows through the activated wormholes after the DNA was removed. The side effect encountered most often in hyper communication and also in human beings are inexplicable electromagnetic fields in the vicinity of the persons concerned. Electronic devices like CD players and the like can be irritated and cease to function for hours. When the electromagnetic field slowly dissipates, the devices function normally again. Many healers and psychics know this effect from their work." (13)

So , we are getting more of an idea how DNA and gene communication actually works . It is through this ' invisible web ' that the communication takes place . So a disruption in this communication network signifies a significant problem .

But how to bring things back into alignment ?

We have already slightly touched on the idea that changing the placement of the regulatory elements within the genome has an effect on the way genes communicate with one another . How would the Garjajev experiment affect cellular and genetic communication ... or is this even addressed ?

What Dr. Garjajev and his colleagues have suggested is that we have a ' living internet ' inside of us , and that " DNA Can Be Influenced and Reprogrammed by Words and Frequencies. " (14) And the method of communication , as mentioned before , is " Living chromosomes function just like solitonic-holographic computers using the endogenous DNA laser radiation." (14) " This means that they managed for example to modulate certain frequency patterns onto a laser ray and with it influenced the DNA frequency and thus the genetic information itself. " (14)

To continue : " Since the basic structure of DNA-alkaline pairs and of language (as explained earlier) are of the same structure, no DNA decoding is necessary. One can simply use words and sentences of the human language!
   This, too, was experimentally proven! Living DNA substance (in living tissue, not in vitro) will always react to language-modulated laser rays and even to radio waves, if the proper frequencies are being used. This finally and scientifically explains why affirmations, autogenous training, hypnosis and the like can have such strong effects on humans and their bodies. It is entirely normal and natural for our DNA to react to language.

Further : " While western researchers cut single genes from the DNA strands and insert them elsewhere, the Russians enthusiastically worked on devices that can influence the cellular metabolism through suitable modulated radio and light frequencies and thus repair genetic defects. " (14)

So , we are not talking about having to move regulatory genetic elements around within the genome , until we have hit upon the correct combination , as the EMBL study would suggest ; but instead we are talking about influencing the way genes communicate with one another and the rest of our bodies , through simple radio and light frequencies .

This brings up a question in my mind ; in that light therapy might go a long way in this type of therapy . I say this because " .... Maurice Cotterell, author of The Mayan Prophecies, Astrogenetics and several other cutting edge books has studied the effects of solar radiation on humans for over twenty years and discovered that genetic mutations are caused through the action of ionising radiations. He has found that X-rays and gamma rays from the Sun are the key factor in genetic leaps of species." (13) Thus , it consists of finding the correct combinations of light (and radio) waves .

Did Garjajev address this , and was he successful ?

Yes ! " His research group succeeded in proving that with this method chromosomes damaged by x-rays for example can be repaired. They even captured information patterns of a particular DNA and transmitted it onto another, thus reprogramming cells to another genome. So they successfully transformed, for example, frog embryos to salamander embryos simply by transmitting the DNA information patterns! " (14)

" This way the entire information was transmitted without any of the side effects or disharmonies encountered when cutting out and re-introducing single genes from the DNA." (14)

This is new science , called ' wave genetics . ' Furthermore :

" Esoteric and spiritual teachers have known for ages that our body is programmable by language, words and thought. This has now been scientifically proven and explained. Of course the frequency has to be correct. And this is why not everybody is equally successful or can do it with always the same strength. The individual person must work on the inner processes and maturity in order to establish a conscious communication with the DNA." (14)

So there it is : " to establish a conscious communication with the DNA." (14) Somewhere along the line , we have lost this ability .

" In their book Vernetzte Intelligenz (Networked Intelligence), Grazyna Gosar and Franz Bludorf explain these connections precisely and clearly. The authors also quote sources presuming that in earlier times humanity had been, just like the animals, very strongly connected to the group consciousness and acted as a group. To develop and experience individuality we humans however had to forget hyper-communication almost completely. "

" Now that we are fairly stable in our individual consciousness, we can create a new form of group consciousness, namely one, in which we attain access to all information via our DNA, without being forced or remotely controlled about what to do with that information." (14)

Wouldn't that be a grand deal ? To be able to access our DNA information with our mind .

Furthermore : " Researchers think that if humans with full individuality would regain group consciousness, they would have a god-like power to create, alter and shape things on Earth! [I am overjoyed. Finally orders from the Universe are scientifically explained!!!] AND humanity is collectively moving toward such a group consciousness of the new kind. Fifty percent of today's children will be problem children as soon as the go to school. " (14)

" The system lumps everyone together and demands adjustment." (14)

So it is a matter of living in a Spiritual state in our high tech world ; and seeing all of this ' high technology ' for what it is . An attempt by man to replicate this group consciousness and group communication .

What type of sounds can heal the DNA ?

" Sound (the word) is the original creational tone. Everything is made up of energy at various frequencies. All things in nature vibrate to sound, light and color. Sound frequencies effect everything about us. The correct vibrational frequencies can be used to heal and balance our bodies. " (14)
" Sound healing is the practice of using sound to realize and correct imbalances in the body. Sound healing works on the belief that the human body is not solid. Rather it is energy that is held together by sound. Any disease therefore indicates that some sound has gone out of tune." (14)

Any disease indicates that some sound has gone gone out of tune . I shall not delve into the technicalities of sound therapy too much , but I would suggest to the reader who is interested in following up on this to follow link number (15) ; and also to research the subject on your own . I would also suggest the subject of music therapy (16) to the seeker who wishes to follow up on the wondrous subject of music , sound therapy and healing .

If one wished to check out some of this sound therapy , I would suggest this list of videos :
https://www.youtube.com/results?search_query=Sound+for+the+healing+of+DNA .

Namaste !

I would like to , for a time , leave these somewhat new and experimental treatments , and devote some time to the prevalent technology for protecting the transplanted beta cell , that of encapsulation .

(1) National Center for Biotechnolgy Information   :   Beta-cell transplantation in type 1 diabetic patients: a work in progress to cure   : http://www.ncbi.nlm.nih.gov/pubmed/20726441

(2) DiabetesNet.com   :   Beta Cell Transplants   : http://www.diabetesnet.com/beta-cell-transplants

(3) Diabetes and the Environment   :   Dioxin and Diabetes / Obesity   : http://www.diabetesandenvironment.org/home/contam/pops/dioxins

(4) Wikipedia   :   2,3,7,8-Tetrachlorodibenzodioxin   : http://en.wikipedia.org/wiki/2,3,7,8-Tetrachlorodibenzodioxin

(5) Wikipedia   :   Aryl hydrocarbon receptor   : http://en.wikipedia.org/wiki/Aryl_hydrocarbon_receptor

(6) A Biotech Primer   :   How Do Cells Communicate?   : http://www.biotechprimerblog.com/2010/08/how-do-cells-communicate.html

(7) Wikipedia   :   Gene expression   : http://en.wikipedia.org/wiki/Gene_expression

(8) Wikipedia   :   Phenotype   :   http://en.wikipedia.org/wiki/Phenotype

(9) Phys org   :   DNA's twisted communication: Genome organization key element for control of gene expression   :   http://phys.org/news/2013-02-dna-genome-key-element-gene.html

(10) Wikipedia   :   Genome   :   http://en.wikipedia.org/wiki/Genome

(11) Wikipedia :   cis-Regulatory element   :   http://en.wikipedia.org/wiki/Cis-Regulatory_element

(12) My Science Academy   :   Scientists Finally Present Evidence on Expanding DNA Strands   :   http://myscienceacademy.org/2013/01/23/scientists-finally-present-evidence-on-expanding-dna-strands/

(13) VirginAir.com   :   Russian DNA Research   : http://www.nrgnair.com/MPT/zdi_tech/DNA.research.htm

(14) bibliotecapleyades.net : DNA's Hyper Communication : The Living Internet Inside of Us :
http://www.bibliotecapleyades.net/ciencia/ciencia_genetica02.htm

(15) Crystalinks.com : Sound Frequencies For Healing :
http://www.crystalinks.com/sound_frequencies.html

(16) Crystalinks.com : Music : http://www.crystalinks.com/music.html

Chapter Eight

Encapsulation and Transplantation

I want to devote a chapter to the most standard methods in use today for curing Type 1 diabetes . And those are encapsulation of beta cells . and their placement into the body ; transplantation of the pancreas itself ; and also , a new method of insulin delivery that is being developed , the Bio-Hub .

Let us begin by taking a look at the Bio-Hub . The DRI (Diabetes Research Institute defines the Bio-Hub as : " a bioengineered "mini organ" that mimics the native pancreas, containing thousands of insulin-producing cells that sense blood sugar and release the precise amount of insulin needed, in real time. " (1)

Apparently the transplantation of the islet (beta) cells themselves have run into a set of problems . These problems include :

Supply – the need for more insulin-producing cells for transplant

Sustainability – the need for the recipient to accept the cells long-term, without the need for anti-rejection drugs

Site – the identification of an optimal site within the body to house the new cells

So , there are rejection problems with the Islet cells , which require the patient to take anti rejection medications , such as Neoral , Cyclosporin and Cellcept . This would be the case for people who have had a straight pancreatic transplant , also - - the need for anti - rejection medications . Make no mistake about it , however . I have , personally , met people who have had the pancreas transplant , and they no longer have the need for insulin injections . I met one young lady who had the dual kidney - pancreas transplant (this is the case with many Type 1 diabetics ; they have the pancreas transplant at the same time as the kidney) ; and she was doing remarkably well with it . But , the problems remain , such as supply , and the need for expensive follow - up medications . So the Diabetes Research Institute (which I shall refer to henceforth as the ' DRI) has developed the Bio-Hub , to address some of these issues .

" The BioHub platform addresses these challenges by drawing on recent developments in bioengineering, immunology, and decades of transplantation expertise.

Prior to their destruction by the immune system in type 1 diabetes, healthy islets thrive inside the pancreas, where they have sufficient oxygen, adequate space, and all the nutrients needed to perform the demanding job of normalizing blood sugar levels.

The BioHub attempts to closely replicate the cells' natural environment and allows scientists to fine tune these cellular needs within the transplant site as never before." (1)

This is a promising sounding procedure . But how successful has it been so far ?

From what I am understanding about the Bio-Hub it is the latest method of encapsulating the beta cells . (More information here (2) And , of course , the Bio-Hub is not now available to patients , the first clinical trials scheduled to start in late 2014 . And there is the continued problem of having to deal with heavy anti - rejection medications (if the trials are successful) .

So how are the trials with the Bio-Hub going ?

In actuality , the trials for the Bio-Hub have not actually started yet (as of March 2015) . But the application form can be found here :

http://www.diabetesresearch.org/diabetes-clinical-trials (which is also a page on the DRI website)

One operation that is going on at the present time is the dual kidney / pancreas transplant . Transplantation of the pancreas is usually done for diabetics who have had kidney failure (not uncommon in Type 1 diabetes) , and are in need of a transplant . Why , exactly , it is done like this is possibly due to mere coincidence , or , possibly more to the point , that the kidney operation is more perfected , and the chances of the two organs surviving together after a dual operation are greater .
But , whatever , the reasons , I have personally seen evidence of the effectiveness of kidney / pancreas transplantation . I , myself , am I kidney transplant recipient , and although I did not receive the pancreas at the same time , I have met at least two different people who did have the kidney / pancreas transplant ; and both were doing remarkably well .

So , if transplantation works so well , what then is the problem with doing this more often ? The answer is , quite simply , one of supply . With the straight kidney transplant , a living donor will suffice , as a person can survive with only one kidney . With the pancreas , however , living donor is not a possibility , as each person has only one , and the pancreas is necessary to maintain the body's blood sugar level . So the supply can become very scarce .

The pancreas from a deceased donor can be stored and transported for up to twenty hours , so this reduces the options somewhat for the pure pancreas transplant , as the recipient must be close enough to the cadaver donor to allow the delivery of the organ within this time period . Thus , the kidney transplant is much more prevalent . The success rates for these various operations are as follows :

" 1) Simultaneous pancreas-kidney transplant. In about 85 percent of people who receive a simultaneous pancreas-kidney transplant, the transplanted pancreas is still functioning after one year. After five years, that rate is about 73 percent."

2) Pancreas-after-kidney transplant. In about 82 percent of people who receive a pancreas-after-kidney transplant, the transplanted pancreas is still functioning after one year. Five years after the transplant, the rate is about 65 percent."

3) Pancreas-only transplant. In about 76 percent of people who receive a pancreas-only transplant, the transplanted pancreas is still functioning after one year. After five years, that rate is about 53 percent." (3)

So , in spite of these success rates , we have a supply problem , and the Bio-Hub has been invented to address this .

Now , let us move on to look at encapsulation of beta cells without the use of the Bio-Hub . What are the latest techniques ?

" These cells are encapsulated in a device that isolates the cells from the host but allows free flow of oxygen, nutrients, and other factors, so that the cells can respond to blood glucose and release hormones like insulin while being protected from the patient's immune system." (4) This is from a company called ' Viacyte , Inc . ' In March of this year (201 %) ViaCyte received the go ahead for clinical tests on it's Stem Cell replacement therapy for Type 1 diabetes . This is only in Canada , however , as stem cell replacement is still a hot button topic here in the U.S. (5)

Stem Cell replacement therapy is surely an exciting , emerging field ; but at the present I wish to get back to encapsulation ; and we shall return to stem cell therapy shortly .

Here is a bit more on the encapsulation , issue , once again from the JDRF :

" New beta cells are created and wrapped in a permeable, protective barrier which is implanted in the body. The new beta cells release insulin when needed while the barrier protects them from being destroyed by the autoimmune attack . " (6)

So , this ' permeable protective barrier ' must be multi - functional ; to not only protect the transplanted beta cells ; but to also allow sufficient interaction with the body . This in order for the cells to be able to sense the blood sugar levels in the body , and to release insulin in order to keep these levels at a normal reading . Encapsulation also increases the number of beta cells available for transplantation , as this is another area that suffers from a lack of supply .

Here is the little video on the JDRF page that explains the function of the permeable protective barrier in regards to the beta cells :

https://www.youtube.com/watch?feature=player_embedded&v=mAjFVtymkmc

Apparently there are spaces in the permeable barrier that let blood in , and also allow insulin to flow out . The video also states that they (the JDRF) is working on finding the best substance , or substances , possibly , to use in the construction of the barrier .

The upside of the encapsulation process is that it does not require the use of heavy doses of anti - rejection medications , as are needed in the straight kidney / pancreas , or pancreas transplant . The ' protective barrier ' itself protects the cells from attack from the body's immune system . At the present time , however , trials with various encapsulation procedure are continuing .

Encapsulation therapy , and testing on humans began through the JDRF in 2014 . More information here : (6)

Let us take a closer look at some of the problems and issues with encapsulation .

" The main motive of cell encapsulation technology is to overcome the existing problem of graft rejection in tissue engineering applications and thus reduce the need for long-term use of immunosuppressive drugs after an organ transplant to control side effects. " (7)

What , exactly , is the encapsulation material ? " The use of the best biomaterial depending on the application is crucial in the development of drug delivery systems and tissue engineering. The polymer alginate is very commonly used due to its early discovery, easy availability and low cost but other materials such as cellulose sulphate, collagen, chitosan, gelatin and agarose have also been employed." (7)

So , let's take a closer look at Alginate (although , as noted above , there are several different materials now in use) .

Alginate : " Several groups have extensively studied several natural and synthetic polymers with the goal of developing the most suitable biomaterial for cell microencapsulation. Extensive work has been done using alginates which are regarded as the most suitable biomaterials for cell microencapsulation due to their abundance, excellent biocompatibility and biodegradability properties. Alginate is a natural polymer which can be extracted from seaweed and bacteria[10] with numerous compositions based on the isolation source. " (7) Alginates do have their problems , however ; which are : " Alginate is not free from all criticism. Some researchers believe that alginates with high-M content could produce an inflammatory response and an abnormal cell growth ; while some have demonstrated that alginate with high-G content lead to an even higher cell overgrowth . " There were also reactions when these alginates were created that produced more proteins .

There have been experiments with other alginate derivatives :

" Researchers have also been able to develop alginate microcapsules with an altered form of alginate with enhanced biocompatibility and higher resistance to osmotic swelling. Another approach to increasing the biocompatibility of the membrane biomaterial is through surface modification of the capsules using peptide and protein molecules which in turn controls the proliferation and rate of differentiation of the encapsulated cells. " (7)

So the search is on-going for the perfect encapsulatiion medium . A different encapsulation medium is used for different functions of the body for which it is used ; i.e. : bone , blood vessel , muscle tissue , cellular , etc . Encapsulation is a complex science , as allowances in the substance used must be made to allow for the entry and exit of molecules . This is called ' microcapsule permeability ' . This is especially important to the diabetic patient and the transplantation of beta cells ; as the beta cell must be allowed to interact with the body , in order to sense blood sugar levels ; and then the beta cell must be able to deliver it's insulin to the body . As different cells , doing different functions , have differing permeability requirements , the study and perfection of this aspect of encapsulation goes on .

Another important aspect of encapsulation is to make sure that the material used is strong enough not to rupture when trasplanted into the body . This is called ' membrane strength ' , or ' mechanical stability ' .

The search for the perfect encpsulation membrane for the beta cells is on-going . This , from the JDRF (Juvenile Diabetes Research Foundation) in 2011 : " Beta cell encapsulation focuses on taking islets, which can be derived from human, pig, or stem-cell sources, and protecting them with a physical barrier. The barrier shields these islets from the immune system, while allowing insulin and nutrients to move freely through. Because the barrier physically hides the islets from the immune system, this approach obviates the need to suppress the entire immune system to stem its attack. " (8)

To continue : " One of the key challenges in the field is selecting or creating a material that protects islets from the immune system while keeping them healthy and functioning properly. To date, researchers have only had short-term success, as islets have been shown to survive for only six months in a variety of animal models using existing materials. "The problem is, we don't know what it is about these materials that is allowing the islets to fail," says Dan Anderson, Ph.D., a biomaterials engineer from the Massachusetts Institute of Technology, who attended the workshop. "So now we are tweaking existing materials and testing them on islets to see what it is about these materials that does and does not work; understanding this will allow us to design better materials." (8)

This search for the perfect encapsulation membrane continued past the 2011 date of the above data , to this February 2015 report . (9) The JDRF was frustrated in attempts to use alginate in higher primates (to protect the beta cells) , after it had worked so well in lower primates . So the JDRF approached " two senior scientists, Daniel Anderson and Robert Langer, and asked them what they needed to take on the complicated problem of protecting beta cells through encapsulation—a holy grail in type 1 diabetes research. " (9)

These two scientists created 800 analogues (types) of alginate in their laboratory , testing each one rigidly to find out which one was the best . After much failure (they were using mice with higher efficiency immune systems than in the previous JDRF studies) they finally came up with three that the immune system seemed not to recognize as a foreign material . Of these three materials they isolated ONE , which became the ' gold standard ' of encapsulation ; this derivative of alginate . (9) With this ' lead material ' (as they also called it) they were able to control the blood sugars in a mouse for a six month period (the length of the experiment) . Dr. Vega is going to publish his paper during the coming year . (The above quoted article (9) is from February 2015) .

The next step is to start clinical tests on primates with the ' Melton Lab beta cells ' , (Melton Lab being where the above research took place) , which should take about two years . And then comes the long clinical trial on human beings . So we are a bit removed yet from true beta cell encapsulation that can be used to cure type 1 diabetes in humans . The Bio-Hub , however , IS ready to start it's testing on humans ; although their search for the perfect material to use goes on also (2) . Right now they are working on two models : (1) a biodegrable material , which " uses the patient's own plasma, the liquid part of the blood that does not contain any cells, together with thrombin, a commonly used, clinical-grade enzyme. When combined, they create a gel-like substance that sticks to the implantation site (the omentum)." (10) (Note : ' the omentum ' is a visceral fold that hangs down from the stomach) . ; and 2) A Bioengineered Scaffold ; which " is a sponge-like disc that is compatible with the human body. These scaffolds are made of only 10 percent silicone. The rest is open space, creating tiny pores that can house thousands of insulin-producing cells of many shapes and sizes." (10)

" The DRI also plans to test this silicone scaffold utilizing the omentum as a transplant site. Researchers are in discussions with the FDA about additional preclinical testing that the regulatory agency has required before approval of that pilot clinical trial in the U.S. can be granted." (10) So , there is work yet to be done on this ' bioengineered ' product .

Let us now move off of the encapsulation side of things , back to the supply side ; and look at stem cell research , as it relates to type 1 diabetes (T1D) .

Stem Cells

The Diabetes Research Institute (DRI) has recently made great strides in the area of stem cell research for Type 1 diabetes , in the discovery of " An area that has sparked great interest is the discovery of stem cells in the "biliary (BILL-ee-air-ee) tree" – a network of drainage ducts that connect the liver and pancreas to the intestine." (

What is the definition of a ' stem cell ' ? According to Wikipedia : " Stem cells are undifferentiated biological cells that can differentiate into specialized cells and can divide (through mitosis) to produce more stem cells." (12)

So what we have with the DRI discovery is a source of undifferentiated cells that are close to the pancreas , and " In the lab, scientists have instructed the biliary cells to mature into islets. These islet structures produced insulin and c-peptide (a component of natural insulin production) in response to glucose. Transplanting these structures into diabetic mice dramatically improved blood sugar control. " (11)

So , it is a matter of manipulating , or ' communicating ' with these undifferentiated cells , and pushing them towards sensing glucose in the blood , and then producing insulin . How is this done , exactly ?

The key is to capture , or find a source (as the DRI has) of non - mature , or ' embryonic ' cells , that can be changed , or ' differentiated ' into whatever function is necessary . The process of differentiating these cells is a rather complex process . Simply put , a ' growth factor ', or signal , must be induced from an already differentiated cell (a cell of the type that you wish the stem cell to differentiate into) . These signals are picked up by receptors in the stem cell , and the stem cell then sends it's own signals to it's nucleus ; and specific DNA sequences are transcribed , and translated into specific cell protein .

Upon reading this , I am struck by the fact of Dr . Garjajev's experiment with communication within the DNA code may be very useful in the future of stem cell research .

Adult stem cells are also now being used in the process . These adult cells are slightly more difficult to get to differentiate , due to their mature state ; but these mature cells have some advantages over the embryonic cells . They " generally do not have the potential for malignancy, are harvested with relative ease, and are available in greater supply." (13) In other words the embryonic (immature) cells are less stable .

Two types of cells are now being studied by the DRI for potential use in Type 1 diabetics : " Cord blood stem cells, as the name implies, are derived from newborn umbilical cord blood. The umbilical cord harbors two types of stem cells: mesenchymal and hematopoietic. Mesenchymal cells potentially could be used as immunomodulators and "helper" cells in the process of endogenous regeneration. Hematopoietic cells are similar to those obtained from the bone marrow, and their primary purpose is to generate the entire array of blood-forming and immune cells. Both cell types are being actively investigated for their potential to reeducate the immune system in Type 1 diabetes. " (13)

So , once again , these cell types are seen for their potential to ' reeducate ' the immune system ; to communicate with the immune system to keep it from attacking and destroying the beta cells , and also for their ability to induce replication of the needed cells .

So .. how are the human tests going at this point ?

In September of 2014 Robert R. Henry, MD, professor of medicine in the Division of Endocrinology and Metabolism at UC San Diego , and his team , " launched the first-ever human Phase I/II clinical trial of a stem cell-derived therapy for patients with Type 1 diabetes." (14) According to Dr. Henry : " We will be implanting specially encapsulated stem cell-derived cells under the skin of patients where it's believed they will mature into pancreatic beta cells able to produce a continuous supply of needed insulin. Previous tests in animals showed promising results. We now need to determine that this approach is safe in people." (14)

The above is a two year study , and will be conducted in two phases .

In another study by Bernat Soria and colleagues indicate that isolated beta cells—those cultured in the absence of the other types of islet cells—are less responsive to changes in glucose concentration than intact islet clusters made up of all islet cell types. Islet cell clusters typically respond to higher-than-normal concentrations of glucose by releasing insulin in two phases: a quick release of high concentrations of insulin and a slower release of lower concentrations of insulin. In this manner the beta cells can fine-tune their response to glucose. " (15) Thus , thee cells are reacting in the same fashion that slow and fast acting insulin injections react to the type 1 diabetic .

There have also been some positive tests using fetal cells , that continue to replicate and differentiate when implanted into mice . (15) However , this study concluded that only undeveloped fetal cells could be induced to differentiate , and that purified cells (already differentiated) could not further proliferate .

So studies with stem cells are ongoing . I have listed a couple of very good sources for information at links (15) and (16) for those who wish to further study the intricacies this expanding field .

I now wish to deviate a bit from transplantation , encapsulation and stem cells ; and in the next chapter take a look at alternatives to insulin therapy for type 1 diabetics .

(1) Diabetes Research Institute   :   DRI BioHub   :
http://www.diabetesresearch.org/BioHub

(2) Diabetes Research Institute   :   BioHub FAQs   :
http://www.diabetesresearch.org/BioHub-FAQ

(3) Mayo Clinic   :   Pancreas transplant   :   Results   :
http://www.mayoclinic.org/tests-procedures/pancreas-
transplant/basics/results/prc-20014239

(4) JDRF San Diego Chapter   :   Research Update – Beta Cell Encapsulation   :
http://sd.jdrf.org/event/research-update-beta-cell-encapsulation/

(5) ViaCyte   :   Stem Cell Trials   :   http://viacyte.com/press-releases/viacyte-
receives-clearance-from-health-canada-for-diabetes-clinical-trial/

(6) JDRF   :   Encapsulation   :   http://jdrf.org/encapsulation/

(7) Wikipedia   :   Cell encapsulation   :
http://en.wikipedia.org/wiki/Cell_encapsulation

(8) JDRF   :   Beta Cell Encapsulation Workshop   :   http://jdrf.org/2011/04/beta-
cell-encapsulation-workshop/

(9) A Sweet Life   :   The Road to a Type 1 Diabetes Cure: Encapsulation   :
http://asweetlife.org/feature/the-road-to-a-type-1-diabetes-cure-encapsulation/

(10) Diabetes Research Institute   :   Scaffolds   :
http://www.diabetesresearch.org/scaffolds

(11) Diabetes Research Institute   :   Stem Cells   :
http://www.diabetesresearch.org/stem-cells

(12) Wikipedia   :   Stem cell   :   http://en.wikipedia.org/wiki/Stem_cell

(13) Diabetes Research Institute : What Are Stem Cells? :
http://www.diabetesresearch.org/Stem-Cells-FAQ#canscientists

(14) News - Medical net : Clinical trial of stem cell-derived therapy for
patients with Type 1 diabetes : http://www.news-medical.net/news/20140911/
Clinical-trial-of-stem-cell-derived-therapy-for-patients-with-Type-1-
diabetes.aspx

(15) National Institute of Health : Stem Cells and Diabetes :
http://stemcells.nih.gov/info/scireport/pages/chapter7.aspx

(16) Diabetes Reserach org : Stem Cells :
http://www.diabetesresearch.org/file/research-publications/2013-Stem-
Cells_Biliary-Tree-Stem-Cells-to-Islets.pdf

Chapter Nine

Alternatives to Insulin

In this chapter I wish to explore some of the alternatives to daily insulin injections for type 1 diabetics . I am aware that most sources will tell you that are **NO** alternatives to daily injections ; but I , personally , am convinced that we are on the verge of a breakthrough in the area of oral medications for type 1 d , much the same as type 2 diabetics control their disease with orla medication .

Therefore , I will start off with the newest oral meds first , and get into the correct food choices that a diabetic should make later .

I would like to start off by looking at a drug called Smylin . Smylin is , however , an injectable form of lowering blood sugars . It is used when a type 1 (or type 2) has become resistant to their insulin ; or , in the case of type 2 , to their oral medications . Smylin is a very powerful tool to lower blood sugar , and it must be used with a good deal of caution . Read more about Smylin at the link below . (1)

Glucocil is a purely oral medication that the jury is still out on . It consists of th following ingredients :

1) Proprietary Mulberry Leaf Extract   :   > Reduces glucose absorption in the intestines

> Reduces glucose production in the liver

> Increases glucose uptake in the cells

> Promotes heart , blood vessel and circulatory health

> Contributes to weight

management

2) **Alpha Lipoic Acid**   : > **Reduces glucose production in the liver**

> **Increases glucose uptake in the cells**

> **Contributes to weight management**

3) **Banaba Leaf Extract**   : > **Reduces glucose production in the liver**

> **Increases glucose uptake in the cells**

4) **Berberries**   : > **Reduces glucose absorption in the intestines**

> **Reduces glucose production in the liver**

> **Increases glucose uptake in the cells**

> **Supports normal blood lipid levels**

> **Promotes heart , blood vessel and circulatory health**

> **Contributes to weight management**

5) **Chromium Picolinate**   : > **Increases glucose uptake in the cells**

6) **Cinnamon Bark Powder** : > **Increases glucose uptake in the cells**

7) **Fish Oil** : > **Promotes heart , blood vessel and circulatory health**

8) **Glymema Sylnestre Extract** : > **Reduces glucose production in the liver**

> **Increases glucose uptake in the cells**

> **Reduces glucose absorption in the intestines**

9) **Insulina (Cissus sycioides) Leaf Extract** : > **Reduces glucose absorption in the intestines**

> **Contributes to weight management**

10) **Wild Grape (Cissus quadrangularis)** : > **Reduces glucose absorption in the intestines**

> **Contributes to weight management**

11) **Vitamins B1 , B6 and B12** : > **Supports sugar metabolism**

12) **Vitamin D** : > **Promotes heart , blood**

Here is the link (and below at (2) where one can purchase Glucocil :
http://www.glucocil.com/landingpage/

The jury is still out on the various benefits of Glucocil . Remembering that this is more a treatment for type 2 diabetics ; and a supplement to insulin for type 1 diabetics .

You can join the discussion at ' Diabetic Connect ' here : (3)

http://www.diabeticconnect.com/diabetes-discussions/general/7080-glucocil

There are clinicl tests with a drug (for type 2 diabetes) called ' Metformin ' ; when used in conjunction with insulin , can be useful in controlling blood sugar in type 1 diabetics . (4)

Of course there is inhaled insulin . A form of insulin inhaler was approved by the FDA on June 28 2014 , called ' Afrezza' . The way that this works is : " Its Dreamboat inhaler delivers a choice of 4u or 8u capsules of insulin powder. Capsules are prepared by loading insulin molecules onto pH-sensitive carrier particles and then drying and placing the powder into plastic capsules. The aerodynamic properties of the particles delivered by the device scatters the insulin powder deep into the lungs when inhaled. Once these pH sensitive particles contact the neutral pH of the lungs, they become a liquid and quickly absorb through the lung membranes. " (5)

There are other inhalers in this field ; such as the GLP - 1 , but the Afrezza appears to be the cream of the crop .

Does inhaled insulin work ? The answer appears to be ' yes ' ; at least according to R. Keith Campbell, RPh. He's a certified diabetes educator and distinguished professor emeritus in diabetes care and pharmacotherapy at Washington State University College of Pharmacy.
He says : " "From the time you inhale it to the time it actually peaks [in the blood] is 15 to 20 minutes," Campbell says. Injected insulin taken before a meal, he says, takes about an hour to peak." (6) And , according to Bruce Bode, MD , a

diabetes specialist in Atlanta who did a clinical trial funded by MannKind Corporation, the drug's developer : " The body also clears Afrezza more quickly than insulin injected at mealtime . Besides its rapid peak, the drug is "pretty much gone in 2 or 3 hours," Bode says. Rapid-acting injected insulins, he says, usually "hang around for about 4 hours. Afrezza is fast in, fast out. It is emulating, in essence, what the pancreas does . " (6)

Afrezza comes in 4 units , and 8 units of insulin powder ; so I would conjecture it is more efficient for users who normally take small doses of insulin . And , at a maximum lasting time of 4 hours , an individual would probably have to take multiple doses a day .

But , it does work . " 8u of Afrezza seems to be delivering the same amount of insulin as an 8u injection of Humalog." (5) Also , see the charts on this page at Diabetes net for a more detailed analysis of the effects of inhaled insulin on the body's blood glucose .

However , most attempts at inhaled insulin have proved to b failures (save for the very latest of attempts ; which , as noted above , still have their drawbacks) .

So the focus has turned back towards the search of an oral solution for type 1 diabetics . In May 2013 an Israeli firm , Oramed Pharmaceuticals , have come up with a candidate ; but this too , is an addition to injections . Information on this particular website is copyrighted , so I will provide the link at (7) .

Here is a breakdown of the classifications of oral medications : (at link (8)

There are other oral meds that have some promise . One is called Teplizumab , which " uses an antibody targeted against a molecule called CD3 to bind to the immune system's T-cells and restrain them from attacking beta cells." (9) This drug is still in it's testing stage , and , if approved , would be available in 2018 . So now we have oral medications being developed that would block the immune system from attacking the beta cells .

As these oral medications are in the testing stage , for the most part , this is something for the future of type 1 diabetic treatment . Could it possibly become a reality ? As researchers make constant strides forward , more is uncovered concerning the treatment of type 1 diabetes . A company in New Zealand , for example , have had patents approved for two new drugs ; ORMD-0801, and an oral GLP-1 , and they are currently in phase 2b testing . (10)

In addition : " The Norwegian University of Science and Technology have developed a new type of capsule,called TAM, or the Trondheim Alginate Microcapsule, which is designed to camouflage the insulin-producing cells to the body's immune system. Ideally this capsule, when fully tweaked, developed and efficacious, will allow nutrients to enter the insulin cells while insulin is transported out. With international cooperation, this project, coined the Chicago Project, intends to find a permanent and functional cure for type 1 diabetes." (10)

" On May 3, 2012 Midatech and Monosol announced positive bioavailability and pharmacokinetic results from a first-in-human Phase 1 clinical study of their Midaform™ Insulin PharmFilm product in a study comprised of 27 healthy volunteers. By delivering monomeric insulin through the use of a rapidly dissolving mucoadhesive film that is placed onto the inside of the cheek, this novel treatment is both fast acting and effective. MidaSol's product showed a faster onset of action compared to subcutaneous insulin. The apparent success of Midaform is yet another exciting treatment on the horizon." (10)

The above report , concerning MidaDorm , brings to mind the insulin patch , which has never really gotten off of the ground . There are continued attempts , however , to create a workable patch , such as the ' The U-Strip ' , by Transdermal Specialties . (11) This is a very promising looking product that delivers the insulin through the use of ultrasonic transmission . They are in the clinical trial phase of this right now , and the product should be available to the general public in about 18 months (from March 2015) . Read more about this product on their website . (11)

But ... to get back to the oral medications for just a moment . Perhaps the greatest proponent of oral medication over injections is Dr. Robert Unger , the professor of internal medicine at The University of Texas Southwestern Medical Center at Dallas . He and his team of researchers have developed an oral medication called ' leptin ' , which works on the theory that it is not the lack of insulin in the human per se that causes diabetes ; but instead it is the lack of glucagon action (' glucagon action ' being glucagon secretion within the pancreas . When insulin is profused , or secreted within the pancreas , the glucagon level goes way down . When insulin secretion stops , glucagon level goes back to normal . The reason that this occurs , says Dr. Unger , is that it is the juxtaposition of alpha cells AND beta cells within the pancreas that controls glucagon levels . (Note : Glucagon is that substance which is secreted in the pancreas that RAISES the level of glucose in our body . It is the exact opposite of insulin . However - - - Dr. Unger found that these glucagon secreting alpha cells were not only located in the pancreas ; but also in the stomach .

Dr. Unger goes on to conjecture that insulin therapy alone is not sufficient to control type 1 diabetes . One gets severe up and down spikes in glucose levels throughout the day . Any type 1 diabetic can attest to this ; particularly with the rise of glucose levels around meal time . Injected insulin simply cannot duplicate the body's natural dispersion of insulin . So giving insulin alone simply cannot control type 1 diabetes (this according to Dr. Unger , of course) . If one were to suppress the amount of insulin that one took (by quite a bit) you would eliminate the low blood sugars (hypoglycemia) ; but you would increase the number of hyperglycemic episodes (high blood sugar) that you would experience . But - - - using another hormone drug ; a bi-hormonal approach , to lower the amount of high glucagon levels would normalize these high – low spikes ; so that the blood sugar records of the type 1 diabetic will then resemble that of the non - diabetic . One can also block the glucagon receptor with a glucagon anti - body in order to eliminate high blood sugar spikes .

And I am supposing that it is the pharmaceitical drug leptin that they have developed that acts as this hormonal agent ; that suppresses high glucagon levels and acts as a stabilzer for blood sugar levels . (12) (13)

But .... does leptin suppress glucagon as well as insulin ? Well , the tests are ongoing ; but Dr. Unger insists that his tests with mice show that it does ; and an individual with type 1 diabetes needs just a small amount of insulin in order to stabilize the blood sugar results . (13)

Another factor concerning leptin is that it possibly acts to restore correct signalling between the brain and the cells , which has been interrupted by the disease of the auto immune system that causes type 1 and type 2 diabetes . More on this studay at link (22) . The cellular communication studies by Dr. Garjajev and his team come to mind as a possible aid in theis testing .

I know what type 1 diabetics are thinking now ! Anything to be free from the pain of the needles ! (ha ha) Apparently not quite yet , however .

Food

I would now like to move onto another phase of glucose control ; and that would be the food itself that we eat . Are there any foods that are capable , by themselves , of lowering our blood sugar level ?

According to WebMD there are several . These include : " Broccoli, green beans and spinach are all vegetables that may help lower blood sugar . " (14) In addition , there are other foods ,such as strawberries , which are a substitute for sweets ; " lean meat, including salmon, is also ideal for controlling blood sugar . Meat is high in protein and has no carbohydrates. It also has chromium, a mineral that helps the body metabolize carbohydrates and helps insulin function correctly. Salmon is extremely good as it contains omega-3 fatty acids . " (14)

" Cinnamon may help lower blood sugar. In a recent German study, it was shown that cinnamon extract can lower blood sugar by more than 10 percent on average. Sugar-free sparkling water is a good choice for people looking to lower blood sugar as it helps to curb the cravings for soda. " (14)

In addition , WebMD recommends : " To lower blood sugar naturally, Laurie Sanchez for LifeScript recommends spacing out carbohydrate intake throughout the day, eating foods that digest slowly without causing blood glucose spikes and getting at least seven hours of sleep each night. Additional tips include maintaining a healthy weight, exercising frequently, practicing yoga and meditation to lower stress levels and eating a high-protein breakfast every morning . " (15)

So what's the deal with cinnamon ? According to WebMD : " Researchers have found supplementation of 1 to 6 grams of cassia cinnamon for 40 days reduces blood sugar by 24
percent and cholesterol by 18 percent, reports WebMD. Other studies of cinnamon's effects on blood sugar levels have been inconclusive . " (16)

" Researchers have speculated that cinnamon reduces blood sugar levels by improving insulin sensitivity in diabetics. This allows the hormone to transport more sugar from the blood into the body's cells, reports WebMD. Study results on the issue provide varying results, with many suggesting that cinnamon is helpful in lowering blood sugar levels, and others demonstrating no effect on blood sugar levels." (16)
" Cinnamon has not been found to have any harmful effects on diabetics, but caution is urged when it is used in addition to other blood sugar-lowering drugs or supplements. Diabetes medications, along with chromium, panax, psyllium, alpha lipoic acid and garlic, must all be considered before adding cinnamon to a diet, warns WebMD. The synergistic effect of these supplements can cause blood sugar to drop too steeply. " (read more at link (16)

So a key for good diabetic control is to stay away from the heavy carbohydrates . What other foods should diabetics avoid ?

This list (according to Diabetic Living online   :

Think Twice Before Eating These Foods

Nachos

Coffee Drinks

Biscuits and Sausage Gravy

Battered Fish Dinners

Fruit Juice Beverages

Deep-Fried Chinese Entrees

Cinnamon Rolls

Restaurant French Fries

Purchased Cookies

Fried Chicken

Purchased Pie

Purchased Smoothies

Processed Lunch Meat

Restaurant Hamburgers

Purchased Doughnuts and Baked Goods

Frozen Meals

Regular Soft Drinks

Purchased Cakes

Flavored Water

Frozen Pizza

Milk Shakes

Restaurant Pizza

(see the details on these foods at link (17) , and why you should avoid them)

So ; let's get back then to some foods that a diabetic (both type 1 and type 2) CAN eat . I have found the following website , the Diabetic Mediterranean diet , to be be full of excellent recipes for diabetic health . The full link can be found at (18) . The thing that strikes me about this diet , is that it seems to be a combination of vegetables , proteins , with some nuts thrown into the recipes . All prepared fresh ; no heavy packaged foods , filled with preservatives and sugar . These recipes look very tasty .

There is also a low - card Mediterranean diet that is listed on this website . This page makes this statement : " Diabetes and prediabetes always involve impaired carbohydrate metabolism; metabolic syndrome and simple excess weight often do, too. Over time, excessive carbohydrate consumption can turn overweight and metabolic syndrome into prediabetes, then type 2 diabetes." (19)

 " The key feature of the Low-Carb Mediterranean Diet is carbohydrate restriction, which directly addresses impaired carbohydrate metabolism naturally." (19)

 So , watch those carbohydrates ! Again , let's look at foods that are high in carbohydrates :

(Note   : These come with an example of a healthy alternative (in parentheses) ; see the website link (20) for the full detailed list that is included) :

1   :   Sugars , Syrups & Sweeteners (Granulated Sugar)

2   :   Candies (Jellied Gumdrops)

3   :   Dried Fruit (Dried Apples)

4   :   Cereals (Frosted Rice Krispies)

 # 5   :   Snacks   (Fat Free Potato Chips)

 # 6   :   Cookies & Cakes (Fat Free Oatmeal Cookies)

 # 7   :   Flour   (Rice Flour)

 # 8   :   Jams & Preserves

 # 9   :   Bread , Toast , Bagels , Pizza (Cinnamon - Raisin Bagel Toasted)

 # 10 :   Potatoes (Hash Browns) (19)

   Once again , the foods listed in parentheses are healthy alternatives . At web link (19) there is a list of foods under each category ; ordered from high in carbohydrates , to lower at the bottom of the lists .

   Once again ; let us reverse directions , and take a look at the perfect diabetic diet :

This from Diabetes .org :

Healthy eating includes eating a wide variety of foods including:

vegetables

whole grains

fruits

non-fat dairy products

beans

lean meats

poultry

fish

(link (20)

So we have an outline for foods that we can operate on here .

I would like to finish this chapter off by once again preenting a list of foods that will directly lower one's blood glucose level :

First , here is a list presented by Termina Nazher ; at E Zine Articles :

http://ezinearticles.com/?10-Foods-That-Lower-Blood-Sugar-Level&id=3604383

(Note : This website prohibits one from quoting any of the articles there in ; but there is some very good information on foods that lower one's blood glucose level here)

This from EHow :

" Foods that lower blood sugar levels should be integrated into a healthy diet as well as kept handy for times when blood sugar elevates. Most foods that lower blood sugar have fat content. According to WebMD, good fats lower insulin resistance. When cells are more sensitive to insulin, blood sugar levels drop. Certain nuts and avocados, by virtue of their fat content, make good snacks and the fat keeps hunger at bay. Other foods, such as sweet potatoes, cinnamon, onions and garlic are either high in fiber, high in antioxidants, or have properties that regulate healthy cholesterol. Healthy cholesterol is an element in diabetic health." (21)

I encourage people to do their own research on finding a diet that just right for you

(1) Smylin : SmylinPen (Pramalintide Acetate) Pen Injector :
http://www.symlin.com/

(2) Glucocil : The Total Blood Sugar Optimizer :
http://www.glucocil.com/landingpage/

(3) Diabetic Connect : Glucocil :
http://www.diabeticconnect.com/diabetes-discussions/general/7080-glucocil

(4) dLife : Type 1 Medications : http://www.dlife.com/diabetes/type-1/diabetes-treatment/type-1-medications

(5) Diabetes net : Will Inhaled Insulin Really Take Your Breath Away? :
http://www.diabetesnet.com/about-
            diabetes/insulin/insulin-delivery/inhaled-insulin

(6) Web MD : Diabetes Health Center : Inhaled Insulin Afrezza: FAQ :
http://www.webmd.com/diabetes/
            news/20140630/inhaled-insulin-afrezza

(7) In-pharmatechnologist.com : Oramed Pharmaceuticals : http://www.in-
pharmatechnologist.com/Drug-Delivery/
            Oral-Insulin-All-Mouth-or-an-Injection-of-Hope-for-Type-1-Diabetes

(8) Diabetes org : What Are My Options? : http://www.diabetes.org/living-
with-diabetes/
            treatment-and-care/medication/oral-medications/what-are-my-options.html

(9) UCSF (University of California at San Francisco) : Type 1 Diabetes Drug
Proves Effective in Clinical Trial :

http://www.ucsf.edu/news/2013/08/107936/type-1-diabetes-drug-proves-effective-clinical-trial

(10) YesPharma : The Possibility of an Oral Insulin Treatment for Type 1 and Type 2 Diabetes is Near :
http://www.yespharma.com/uploads/files/Oral%20Insulin.pdf

(11) Transdermal Specialties : The U-Strip – Insulin Patch :
http://www.transdermalspecialties.com/u-strip-patch.html

(12) Diabetic Mediterranean Diet : Dr. Roger Unger and his Glucagon-Centric Diabetes Model :
http://diabeticmediterraneandiet.com/2015/01/10/dr-roger-unger-and-his-glucagon-centric-diabetes-model/

(13) Science business eXchange : Leptin instead of insulin :
http://www.nature.com/scibx/journal/v3/n10/full/scibx.2010.296.html

(14) WebMD : What vegetables lower blood sugar? :
http://www.ask.com/health/vegetables-lower-blood-sugar-30c83c2e50c0c3b0?ad=semD&an=msn_s&am=exact&o=5171&qo=boostResultOnSERP

(15) Ask dot com : How do you lower your blood sugar naturally? :
http://www.ask.com/health/lower-blood-sugar-naturally-64b917d5dfb763a0?qo=questionPageSimilarContent

(16) Ask dot com : How can you lower blood sugar levels with cinnamon? :
http://www.ask.com/health/can-lower-blood-sugar-levels-cinnamon-5617099675914309?qo=questionPageSimilarContent

(17) Diabetic Living Online : 22 Foods to Avoid with Diabetes :
http://www.diabeticlivingonline.com/food-to-eat/nutrition/22-foods-to-avoid-diabetes

(18) Diabetic Mediterranean Diet : Home Page :
http://diabeticmediterraneandiet.com/

(19) Health Aliciousness : Top 10 Foods Highest in Carbohydrates (To Limit or Avoid) : http://www.healthaliciousness.com/articles/foods-highest-in-carbohydrates.php#healthy-high-carb-foods

(20) Diabetes org   :   What is a Diabetes Meal Plan?   :
http://www.diabetes.org/food-and-fitness/food/planning-meals/diabetes-meal-plans-
and-a-healthy-diet.html

(21) E How   :   Foods That Lower Blood Sugar   :
http://www.ehow.com/about_4680699_foods-that-lower-blood-sugar.html

(22) SciBx   :   Leptin Instead of Insulin   :
http://www.nature.com/scibx/journal/v3/n10/full/scibx.2010.296.html

Chapter Ten

Emergencies without Insulin

I thought that I would dedicate a short chapter to a situation where there might be some type of an emergency situation that has taken place ; i.e. - a war , or some other type of survivalist situation , where one might find the pharmacy outlets closed , and the type 1 diabetic might find himself / herself without their normal supply of insulin .

Is it possible to make , or created insulin on , say , your stove ? With a simple chemistry set ?

The short answer is ' no ' ; it is not possible to create usable insulin that easily . The best way to prepare for an emergency situation is to stash as much insulin as you are able to do so . A supply of , say , twelve (12) vials of some type of NPH insulin ought to give you one years supply of insulin . After that ... well . I will try to get into that scenario a bit later in the chapter .

But , that being said , the only way being to buy and stash insulin (in your refrigerator ; or , if there is no refrigeration available , in the coolest and darkest part of your home) , the facts ARE that some people have made insulin at home - - - but only when faced with the direst of circumstance .

I thought that I might start out with how insulin was first discovered .

" In 1920, Dr. Frederick Banting wanted to make a pancreatic extract, which he hoped would have anti-diabetic qualities. In 1921, at the University of Toronto, Canada, along with medical student Charles Best, they managed to make the pancreatic extract." (1)

" Their method involved tying a string around the pancrease duct. When examined several weeks later, the pancreatic digestive cells had died and been absorbed by the immune system. The process left behind thousands of islets. They isolated the extracts from the islets and produced isletin. What they called isletin became known as insulin. " (1)

" Banting and Best managed to test this extract on dogs that had diabetes. They discovered insulin. In fact, they managed to keep a dog, that had had its pancreas taken out, alive throughout the whole summer by administering it the extract (which was, in fact, insulin). The extract regulated the dogs blood sugar levels." (1)

" At this point, Professor J. MacLeod, who had placed the laboratory at their disposal, said he wanted to see a re-run of the whole trial. After doing so he decided to get his whole research team to work on the production and purification of insulin." (1)

" J.B. Collip joined the scientific team, which now consisted of Banting, Best, Collip and MecLeod. They managed to produce enough insulin, in a pure enough form, to be able to test it on patients." (1)

" In 1922 the insulin was tested on Leonard Thompson, a 14-year-old diabetes patient who lay dying at the Toronto General Hospital. He was given an insulin injection. At first he suffered a severe allergic reaction and further injections were cancelled. The scientists worked hard on improving the extract and then a second dose of injections were administered on Thompson. The results were spectacular." (1)

The rest of ther story was equally as spectacular . Once the extract was purified , the scientific team went from ward to ward at the Toronto Hospital , injecting diabetic patients with the new - found insulin . This created many joyous moments for the patients and their families , as diabetic patients awoke from their diabetic ketoacidosis comas .

The key point here is how the insulin itself was extracted form the dog pancreas . In Shanghai , China , during World War 2 , we find a similar case where insulin was created by a young lady , Eva Saxl and her husband Victor , after the Japanese had occupied the city of Shanghai , and closed all of the pharmacies .
Eva , the diabetic , and her husband were lent some laboratory equipment , and decided to attempt to extract the insulin from the pancreas of water buffalo , who were common in the area . They DID suceed , after following the methods that Banting and Best outlined in the book ' Beckman's Internal Medicine ' .

Banting's thinking on this was :

" Late in the nineteenth century, scientists had realized there was a connection between the pancreas and diabetes. The connection was further narrowed down to the islets of Langerhans, a part of the pancreas. From 1910 to 1920, Oscar Minkowski and others tried unsuccessfully to find and extract the active ingredient from the islets of Langerhans. While reading a paper on the subject in 1920, Banting had an inspiration. He realized that the pancreas' digestive juice was destroying the islets of Langerhans hormone before it could be isolated. If he could stop the pancreas from working, but keep the islets of Langerhans going, he should be able to find the stuff! He presented this idea to Macleod, who at first scoffed at it. Banting badgered him until finally Macleod gave him lab space, 10 experimental dogs, and a medical student assistant." (2)

" In May, 1921, as Macleod took off for a holiday in his native Scotland, Banting and his assistant Charles Best began their experiments. By August they had the first conclusive results: when they gave the material extracted from the islets of Langerhans (called "insulin," from the Latin for "island") to diabetic dogs, their abnormally high blood sugars were lowered. Macleod, back from holiday, was still skeptical of the results and asked them to repeat the experiment several more times. They did, finding the results the same, but with problems due to the varying purity of their insulin extract." (2)

" Macleod assigned chemist James Bertram Collip to the group to help with the purification. Within six weeks, he felt confident enough of the insulin he had isolated to try it on a human for the first time: a 14-year-old boy dying of diabetes." (2)

So let us go back , and study this process closely :

" In October 1920, Canadian Frederick Banting concluded that it was the very digestive secretions that Minkowski had originally studied that were breaking down the islet secretion(s), thereby making it impossible to extract successfully. He jotted a note to himself: "Ligate pancreatic ducts of the dog. Keep dogs alive till acini degenerate leaving islets. Try to isolate internal secretion of these and relieve glycosurea." (2)

The idea was the pancreas's Internal secretion, which, it was supposed, regulates sugar in the bloodstream, might hold the key to the treatment of diabetes. A surgeon by training, Banting knew certain arteries could be tied off that would lead to atrophy of most of the pancreas, while leaving the islets of Langerhans intact. He theorized a relatively pure extract could be made from the islets once most of the rest of the pancreas was gone." (2)

So Banting tied off the arteries leading to the pancreas , leading it to atrophy ; but leaving the Islets of Langerhans (and it's extract - - insulin , intact) .

To be more specific : " Banting's method was to tie a ligature around the pancreatic duct; when examined several weeks later, the pancreatic digestive cells had died and been absorbed by the immune system, leaving thousands of islets. They then isolated an extract from these islets, producing what they called "isletin" (what we now know as insulin) (3)

So , how were the first batches of insulin purified ?

Once Banting and Best had succeeded in isolating insulin (first from the laboratory dogs , and then from a fetal calf that NcCleod had supplied to them) a biochemist by the name of James Collip was called in by McCleod to oversee the purification process . After all , the first batches of insulin that had been tried on human patients had caused these patients to have severe allergic reactions due to the impurities present .

So , how was the purification done ?

Well , there are a number of purification methods that biochemists use (4) ; but the most likely method that Mr. Collip used would have been that of a method called ' downstream processing ' , which is used extensively in the purification of biosynthetic products, particularly pharmaceuticals .

The steps in downstream processing are as follows :

1) " Removal of insolubles is the first step and involves the capture of the product as a solute in a particulate-free liquid, for example the separation of cells, cell debris or other particulate matter from fermentation broth containing an antibiotic

2) Product isolation is the removal of those components whose properties vary markedly from that of the desired product.

3) Product purification is done to separate those contaminants that resemble the product very closely in physical and chemical properties. Consequently steps in this stage are expensive to carry out and require sensitive and sophisticated equipment. This stage contributes a significant fraction of the entire downstream processing expenditure. Examples of operations include affinity, size exclusion, reversed phase chromatography, crystallization and fractional precipitation.

4) Product polishing describes the final processing steps which end with packaging of the product in a form that is stable, easily transportable and convenient. Crystallization, desiccation, lyophilization and spray drying are typical unit operations." (4)

So it can be definitely seen that this purification process is a complicated step , and must be done under a microscope .

But how then did Eva Saxl , and her crew in Shanghai , China accomplish this task ? Well , apparently Eva and Victor Saxl did very little of the above purification

. Eva simply went out on a limb , and tried the insulin on herself .And it worked ! ... with very little side effects . It not only worked , but Eva and Victor Saxl were able to produce enough of their ' homemade ' insulin to keep several hundred type 1 diabetics alive in Shanghaai during the war . (Read more on their story at link (5)

This I really don't recommend . At least use some simple purification method , such as heating / boiling etc .

Here is a link to the Eva Saxl video on YouTube   :
https://www.youtube.com/watch?v=G4pNApoNtGI

Let's review the entire process - - - this time from a different source - ' Doom and Bloom ' ; a survivalist Blog (6)

Here it is :

" The present method of preparation is as follows. The beef or pork pancreas is finely minced in a larger grinder and the minced material is then treated with 5 c.c. of concentrated sulphuric acid, appropriately diluted, per pound of glands. The mixture is stirred for a period of three or four hours and 95% alcohol is added until the concentration of alcohol is 60% to 70%. Two extractions of the glands are made. The solid material is then partially removed by centrifuging the mixture and the solution is further clarified by filtering through paper. The filtrate is practically neutralized with Sodium Hydroxide. The clear filtrate is concentrated in vacuo to about 1/15 of its original volume. " (6)

Continued : " The concentrate is then heated to 50o degrees Centigrade, which results in the separation of lipoid and other materials, which are removed by filtration. Ammonium sulphate (37 grams. per 100 c.c.) is then added to the concentrate and a protein material containing all the Insulin floats to the top of the liquid. The precipitate is skimmed off and dissolved in hot acid alcohol. When the precipitate has completely dissolved, 10 volumes of warm alcohol are added. The solution is then neutralized with NaOH and cooled to room temperature, and kept in a refrigerator at 5oC for two days. At the end of this time the dark coloured supernatant alcohol is decanted off.
   The alcohol contains practically no potency. The precipitate is dried in vacuo to remove all trace of the alcohol. It is then dissolved in acid water, in which it is readily soluble.
   The solution is made alkaline with NaOH to PH 7.3 to 7.5. At this alkalinity a dark coloured precipitate settles out, and is immediately centrifuged off. This

precipitate is washed once or twice with alkaline water of PH 9.0 and the washings are added to the main liquid." (6)

Continued : " It is important that this process be carried out fairly quickly as Insulin is destroyed in alkaline solution. The acidity is adjusted to PH 5.0 and a white precipitate eadily settles out. Tricresol is added to a concentration of 0.3% in order to assist in the isoelectric precipitation and to act as a preservative. After standing one week in the ice chest the supernatant liquid is decanted off and the resultant liquid is removed by centrifuging. The precipitate is then dissolved in a small quantity of acid water." (6)

And the conclusion : " A second isoelectric precipitation is carried out by adjusting the acidity to a PH of approximately 5.0. After standing over night the resultant precipitate is removed by centrifuging. The precipitate, which contains the active principle in a comparatively pure form, is dissolved in acid water and the hydrogen ion concentration adjusted to PH 2.5. The material is carefully tested to determine the potency and is then diluted to the desired strength of 10, 20, 40 or 80 units per c.c. Tricresol is added to secure a concentration of 0.1 percent. Sufficient sodium chloride is added to make the solution isotonic. The Insulin solution is passed through a Mandler filter. After passing through the filter the Insulin is retested carefully to determine its potency. There is practically no loss in berkefelding. The tested Insulin is poured into sterile glass vials with aseptic precautions and the sterility of the final product thoroughly tested by approved methods." (6)

So here we have the exact process and materials to turn out and purify a vial of usable insulin . As you can see , it is not an easy process .

So ; what about making synthetic insulin in a home laboratory ? In today's high tech world , surely this is possible , correct ?
In a short answer - - no , it is not . Recombitant DNA insulin is a very complex process that relies on basic recombitant DNA techniques and an understanding of the insulin gene.

There are two basic methods involved , and here they are :

" 1) The insulin gene is a protein consisting of two separate chains of amino acids, an A above a B chain, that are held together with bonds. Amino acids are the basic units that build all proteins. The insulin A chain consists of 21 amino acids and the B chain has 30.

2) Before becoming an active insulin protein, insulin is first produced as preproinsulin. This is one single long protein chain with the A and B chains not yet separated, a section in the middle linking the chains together and a signal sequence at one end telling the protein when to start secreting outside the cell. After preproinsulin, the chain evolves into proinsulin, still a single chain but without the signaling sequence. Then comes the active protein insulin, the protein without the section linking the A and B chains. At each step, the protein needs specific enzymes (proteins that carry out chemical reactions) to produce the next form of insulin." (7)

So we can see that making human recombitant DNA insulin is an extremely complex process , which requires knowledge of microbiology and DNA manipulation . For our purposes here I will not delve into it any further , as it will be beyond the reach of all but the most specialized of laboratories . But , for those wishing to explore this new technology further , I suggest following the link at (7) to further study how human insulin is made .

For those interested in how the first DNA charts of insulin were made , I would suggest studying the work of Frederick Sanger , at his biography at Wikipedia (8) This provides a somewhat simpler view of the process ; but still , I would deem it too difficult for the layman at home to perform .

Now , at Eli Lilly and Company , the fermenting of insulin (adding the insulin gene to a safe strain of the E coli bacteria allows them to manufacture a vast amount of insulin at their production plant in Indianapolis , Indiana , U.S. Researchers have taken the basic insulin gene , and tweaked it slightly here and there , to come up with multiple off - shoots of insulin ; such as : insulin lispro (Humalog) and insulin glargine (Lantus). (9)

The basic process at Eli Lilly goes something like this :

" At Lilly, insulin-making E. coli is grown in 50,000-liter tanks called fermentors. There are more than 5,000 tanks on site. According to Lilly, a batch of insulin from one fermentor could produce a year's supply of insulin for thousands of people. "Our facilities are designed to produce insulin crystals in multiple metric-ton quantities," Walsh says. " (9)

" The E. coli have humble beginnings. Small tubes of the bacteria have been stored in a freezer at minus 70 degrees Celsius (minus 94 degrees Fahrenheit) for decades. Lilly produced a granddaddy batch of E. coli, now referred to as the "master cell bank," sometime in the 1980s. It has gone on to seed every batch of Humulin to this day. Whenever Lilly wants a fresh stash of Humulin, workers go to the freezer, pull out a tube from the master cell bank, thaw it out, and stimulate the bacteria to grow." (9)

   The E coli bacteria having already been combined with the various offshoots of the insulin gene , of course . (see the Frederick Sanger link above to see exactly how this gene was identified and isolated .)

   Further more : " Starting with a mere half gram of bacteria, the microorganisms begin to replicate prodigiously, doubling their numbers every 20 minutes or so. Once a tube gets too crowded, the bacteria are moved into larger and larger domiciles, from flask to bigger flask and from tank to bigger tank. All the microorganisms need to flourish is a source of water, sugar, salt, and nitrogen, which their handlers generously supply. In addition, the bacterial broth contains an additive that helps keep any contaminating microorganisms at bay, says Walsh. Typically, the E. coli are engineered to be resistant to a particular antibiotic, such as ampicillin, so that adding ampicillin to a broth will kill off everything but the prized protein producers. After several days of reproduction, the bacteria are
now ready to start their real job—making insulin. " (9)

   So , in other words , the type 1 diabetics of Indianapolis , Indiana ought to be semi - safe in a survivalist scenario ; as would those near other Eli Lilly production facilities in France , India and China (as Eli Lilly is in the process of expanding it's overseas production) , as well as a vast expansion to it's Indianapolis plant .

   Other insulin producers in the United Sates (and nearby neighbors) include the following :

1) Eli Lilly   :   Indianapolis , Indiana , U.S.

2) Novo Nordisk   :   West Princeton , New Jersey , U.S. website :
http://www.novonordisk-us.com/documents/home_page/document/index.asp

3) Sem BioSys   (this is a Canadian Company) :   Calgary , Alberta

4) GlaxoSmithKline   :   Middlesex , United Kingdom   :   see their website for
U.S. locations   :   http://us.gsk.com/

5) Sanofi (Latus Insulin) :   manufacturing   :   Kansas City & St.Louis , MO. ,
U.S. :   website : http://en.sanofi.com/ (10)

Here is a list of pure drug companies that deal in insulin here in the United States :

Abbot Laboratories - 847-937-6100
Amylin Pharmaceuticals - 858-552-2200
Aventis Pharmaceuticals - 800-633-1610
Bayer Diagnostics - 800-348-8100
Becton-Dickinson - 201-847-6800
Bristol-Myers Squibb - 212-546-4000
Eli Lilly & Company - 317-276-9624
GlaxoSmithKline - 888-825-5249
LifeScan - 800-227-8862
Novartis Pharmaceuticals - 888-669-6682
Novo Nordisk - 800-727-6500
Pfizer - 212-733-2323
SmithKline Beecham
Takeda Pharmaceuticals - 877-582-5332
Valeant - 800-548-5100 (11)

Note : One would have to call these numbers , or visit the website to find the location of the company that might be closest to you . This link also contains the names of companies that manufacture other diabetic products ; such as monitors , test strips , insulin pumps , etc .

Companies in other parts of the world :

Not unsurprisingly , China has the greatest number of insulin manufacturers (due to the large population , of course) .

Here is a link to some of the larger complanies involved :

http://www.alibaba.com/insulin-manufacturers.html

And here is another good list of worldwide producers of insulin and diabetic products :

http://www.themedica.com/drug/anti-diabetic-drug/insulin.html

In conclusion of this chapter , I must say that manufacturing your own insulin will prove to be a daunting task for most people ; although it has been done in the past . I , personally , believe that the best course would be to purchase extra insulin (a bottle or two) each month as you buy your regular supply of insulin ; then store this extra insulin in the coolest , darkest portion of your home . In lieu of that , if you have no insulin , the best course might be to follow a VERY strict diet , and possibly take an over the counter medication like glucagon . I have listed some foods in previous chapters certain foods (such as okra , garlic , cinnamon , grapefruit) which have glucose lowering qualities . Get exercise , meditate , drink plenty of water .

Other than that all I can say is good luck and God bless you

(1) Medical News Today : Discovery of Insulin :
http://www.medicalnewstoday.com/info/diabetes/discoveryofinsulin.php

(2) PBS org : Banting and Best isolate Insulin 1922 :
http://www.pbs.org/wgbh/aso/databank/entries/dm22in.html

(3) Wikipedia : Insulin : Extraction and Purification :
http://en.wikipedia.org/wiki/Insulin

(4) Wikipedia : List of purification methods in chemistry :
http://en.wikipedia.org/wiki/List_of_purification_methods_in_chemistry

 (5) A Sweet Life : Eva Saxl: Surviving as a Diabetic During World War II :
http://asweetlife.org/beccak/blogs/type-1-blogs/eva-saxl-surviving-as-a-
    diabetic-during-world-war-ii/6784/

(6) Doom and Bloom : How To Make Insulin :
http://www.doomandbloom.net/how-to-make-insulin/

(7) Made how dot com : Diabetes Treatment :
http://www.madehow.com/Volume-7/Insulin.html

(8) Wikipedia : Frederick Sanger :
http://en.wikipedia.org/wiki/Frederick_Sanger

(9) Diabetes Forecast : Making Insulin :
http://www.diabetesforecast.org/2013/jul/making-insulin.html

(10) Jaz D Life Science : Insulin :
http://www.jazdlifesciences.com/pharmatech/leaf/BioPharma/Hormones/Insulin.ht
m

(11) Diabetes net : Diabetes Companies :
https://www.diabetesnet.com/diabetes-resources/diabetes-companies

Chapter 11

How to Prove YOUR diabetes was cused by GE PCBS

In this chapter I wish to get into some of the legal ramifications for people and residences who have been exposed to high PCB conditions .

Residential property damage has been much easier to prove ; as there are any number of comprehensive soil tests that can be performed that will tell the researcher exactly how much PCB is in the soil , and , in the case of areas that have had General Electric (or other manufacturers) in the area , a fairly accurate assumption can be made as to how PCBs 9 at certain levels) got into the soil .

With people , it is somewhat of a different story ; as some foods and food packaging contain polychlorinated biphenyls and it's derivatives . So , the trick is to prove that it exists in your system at such a level , and a certain location (a location that has proven industrial PCB poisoning) that it could only have been caused by GE or Monsanto (the manufacturer of PCBs) .

But how to do that ?

The most extensive physical testing of residents has taken place in Anniston , Alabama . As one will recall , I mentioned earlier about the PCB poisoning in Anniston . (1) So , let's take a look at some of the findings .

" A study of more than 1,000 east Alabama residents who live amid one of the world's worst pockets of PCBs contamination found health concerns including heart problems and diabetes that researchers said could be linked to the chemical." (2)

This was from a study concluded in 2008 concerning this small city in Alabama that had been the home of a Monsanto plant (a subsidiary of GE) from 1935 to 1975 . This Monsanto facility was dedicated to the mass production of polychlorinated bipheyls . During this time they " flushed tens of thousands of pounds of PCBs into nearby creeks and buried millions more pounds in a hillside landfill." (3) Much the same as in the surrounding areas near the General Electric in Pittsfield , Mass .

For a long time , after the cessation of the dumping of PCBs by Monsanto , residents did not link the " cancer, learning disabilities, increased asthma rates and reproductive deformities experienced by those living closest to the Monsanto plant were not considered linked to environmental contamination." (3)

" People in West Anniston, which was closest to the Monsanto plant had cancer, kidney and liver problems but it wasn't until local fishermen began pulling deformed fish out of local creeks that anyone blamed Monsanto. People in Anniston didn't want to accept it but in 1995 soil test showed PCB levels higher than they had ever been recorded. Pediatricians at the local clinic reported babies with extremely rare birth defects that could not be explained. "We lead the state in birth defects", reports Dr. Angela Martin." (3)

" In 2000, the U.S. Department of Health and Human services conducted a study on the people that lived closest to the Monsanto plant. Participants included 37 children and 43 adults. In the adults, the PCB level ranged from non-detected to 210 parts per billion. In the children the PCB levels ranged from non-detected to 4.6 parts per billion. The usual PCB load for an adult is 2 parts per billion." (3) (U.S. Department of Health &Human Services)

" In 1993, the Alabama Power Company was breaking ground on land obtained from Monsanto. During this time, a landfill was discovered where Monsanto had illegally dumped PCB contamination. This turned out to be one of two unlined PCB dumps that Monsanto had not reported. It was shortly after this discovery that residents began to attribute their increased health problems to environmental pollutants. In 1995, the congregation of a local west Anniston church was approached by Monsanto management and offered a large sum of money for their property. It was discovered that Monsanto knew that the area in which the church was located was heavily contaminated and needed to be destroyed so that cleanup could take place." (3)

" Individual claims took a step forward in February, 2002 when a jury ruled that Monsanto was in fact responsible for polluting the town of Anniston with tons of toxic PCBs. The ruling was a major victory for residents of the town, who have sued the company over damages to their property, their health and to their emotional well being. The company was found liable on six counts: negligence, nuisance, suppression of the truth, trespass, wantonness and outrage . " (3)

In 2003 Monsanto and Solutia , Inc . settled with more than 20.000 residents for $ 700 million dollars . (4)

But what about the particulars ? As part of the above settlement Monsanto and Solutia agreed to help set up study panels , and to help provide medications for the people of Anniston , Alabama . One such study group was the Anniston Environmental Health Research Consortium . And here are some of the findings of their report :

(Note : In this report I am going to focus in particular on diabetes ; although PCBs have been linked to cancer , learning disabilities and other birth defects and deformations) .

" Conclusions: We observed significant associations between elevated PCB levels and diabetes mostly due to associations in women and in individuals < 55 years of age." (5)

One must also keep in mind that at the time of this study many of the homeowners of West Anniston had moved ; due to the condemnation of their property , or over concerns about their health ; so we are dealing here with a small sample size . Of the 1,110 individuals who were still here , " 774 agreed to a clinic visit, which included measurements—by standard protocol—of height, weight, waist circumference, and blood pressure, as well as a review of current medications. Fasting blood was obtained for analyses of glucose and lipids (Jacksonville Medical Center, Jacksonville, AL), the major 35 ortho-substituted PCB congeners, and 13 pesticides and herbicides [Division of Laboratory Sciences at the Centers for Disease Control and Prevention's (CDC) National Center for Environmental Health, Atlanta, GA]. The PCB congeners and pesticides were measured in serum using high-resolution gas chromatography /isotope-dilution high-resolution mass spectrometry (SjÖdin et al. 2004). Serum total lipids were calculated with the enzymatic summation method using triglyceride and total cholesterol measurements. " (5)

The results are :

(Note : there are very complicated notes on the procedures for collecting and analyzing this data . For any interested parties , I would suggest going to link (5) , and studying this data .)

Once again , the results are these :

" Relationships between PCB levels and diabetes status are shown in Tables 2 and 3.3. (see link (5) Participants with diabetes had significantly higher age-adjusted GM PCB levels than did those with prediabetes or normoglycemia (Table 2); age-adjusted arithmetic means did not differ significantly by diabetes status. Age-adjusted GM PCB levels also were significantly higher for both females and males classed as having diabetes compared with those with prediabetes or normoglycemia. Although mean PCB levels for whites and nonwhites were higher in those with diabetes than in those without diabetes, the differences were not statistically significant. GM PCB levels were similar for males and females, whereas nonwhites had significantly higher PCB levels than did whites. In general, PCB levels were higher in participants of more than 55 years of age than in those < 55 years of age (Table 3). Within specific age strata, PCB levels remained similar across BMI categories for those over 55 years of age. PCB levels showed inverse associations with BMI in younger individuals with diabetes, and in those with prediabetes, although the trends were not statistically significant." (5)

So , there IS a pretty definitive test to see if your diabetes (or other illness , for that matter) was caused by PCBs . This study was done between October 2005–April 2007 ; and dumping of PCBs was stopped in Anniston in 1977 (although much of the PCBs did remain in the area ; due to the continued presence of polluted land , and undiscovered chemical dumps ; in addition to the long periods of time that it takes the PCBs to break down . So , in other words , even years after the fact , in areas that were heavily polluted by polychlorinated biphenyls , such as Anniston , AL . ; Pittsfield , MA ; Schenectady , NY and portions of the Hudson River Valley , there ARE tests that can be done to determine if your diabetes / illness was caused by exposure to PCBs .

The people to get in touch with concerning these tests would be :

1) Division of Laboratory Sciences at the Centers for Disease Control and Prevention (CDC) :   tel # 1 800 232 4636

2) National Center for Ennvironmental Health (NCEH) :   same tel number as bove :   1 800 232 4636

3) Jacksonville Medical Center :   Jacksonville , AL :   tel number :   256-435-4970

And ,

(4) Anniston Environmental Health Research Consortium :   contact through the tel number listed above (for the governmental agencies that they are a part of) at 1800 232 4636
      The members of the consortium are listed at :
http://www.atsdr.cdc.gov/sites/anniston_community_health_survey/anniston_enviro nmental_health_consortium_members.html

I must also add that if you are located near a large hospital or medical facility , they should be able to perform a test for PCB poisoning . I bel;ieve that the code for this test is ' PCBZ ' .

Here is the World Health Organization report on PCBs :

http://www.inchem.org/documents/hsg/hsg/hsg68.htm#SubSectionNumber:2.5.1

Here is the W.H.O.'s findings on the hazards of PCBs :

" 5.1 Hazards

PCBs and PCTs are very resistant to degradation and hence very persistent in the environment. Because they are very soluble in lipids they bioaccumulate, especially in the fatty tissues of all living organisms, and biomagnify in the higher trophic levels of the food-chain.

Although their acute toxicity is relatively low, bioaccumulation and biomagnification may lead to lethal effects, especially at the highest trophic levels. Reduced growth and reproduction may affect populations." (6)

There have been many lawsuits involving PCBs ; 61 of which involved personal injury (between 1980 - 2001) (7) This particular link (7) has much information to offer on the possible PCB lawsuit .

Although these smaller , personal injury lawsuits DO favor the defendant (Monsanto , GE) in most cases ; there have been successful cases involving personal injury and PCBs ; most notably the Monsanto settlement in the Anniston , Alabama poisoning case ; where Monsanto chose to settle both the personal injury cases (more than 1,000 cases were involved) , and the civil suits involving the destruction of property and the clean up . The story of this personal injury settlement can be found here . (8)

There were certain extenuating circumstances in the above case , as Monsanto chose to settle both types of cases due to the length of the court proceedings . Most personal injury cases involving PCBs are a very difficult climb ; as there are many different factors that must be proven : i.e. that the PCBs caused your particular disease . Thus I have posted the blood sampling information above .

To those who wish to pursue such litigation nonetheless , I am listed several attorneys whom are willing to take on such cases .

These include :

1) Wieitz and Luxenberg   :   http://www.weitzlux.com/thankyou.html   :   1 800 476 6070

2) Environmental Litigation Group P.C.   :   http://elgweb.squarespace.com/pcbs/ :   1 800 749 9200

3) Allen Stewart Attorney   : http://www.allenstewart.com/practice_areas/nhl/nhl-settlement.php   :   1 866 440 2460

4) Roetzel and Andress   :   http://www.ralaw.com/practice.cfm?id=13   : Roetzel and Andress have multiple offices located around the U.S. ; all of the locations can be found on this page :

   http://www.ralaw.com/offices.cfm   :   The New York office telephone number is   :   212 803 8160

I wish anyone who is thinking of pursuing this type of litigation good luck .

(1) EPA dot gov   :   Anniston PCB Site   :
http://www.epa.gov/region4/superfund/sites/npl/alabama/anpcbstal.html

(2) GM Watch   :   Poisoned By Monsanto's PCBs   :   http://gmwatch.org/latest-listing/47-2008/10487-poisoned-by-monsantos-pcbs

(3) Commonwealth dot org   :   PCB Pollution in Anniston, Alabama   :
http://www.commonweal.org/downloads/programs/Anniston_AL_PCB.pdf

(4) NY Times   : $700 Million Settlement in Alabama PCB Lawsuit   :
http://www.nytimes.com/2003/08/21/business/700-million-settlement-in-alabama-pcb-lawsuit.html

(5) Environmental Health Perspectives   :   Polychlorinated Biphenyl (PCB)
Exposure and Diabetes: Results from the Anniston Community Health Survey
   :   http://www.ncbi.nlm.nih.gov/pmc/articles/PMC3346783/

(6) Inchem   :   POLYCHLORINATED BIPHENYLS (PCBs) AND
POLYCHLORINATED TERPHENYLS (PCTs) HEALTH AND SAFETY GUIDE   :

http://www.inchem.org/documents/hsg/hsg/hsg68.htm#SubSectionNumber:2.5.1

(7) Beasley Allen dot com   :   MONSANTO, PCB'S AND FUTURE TOXIC TORTS
:   Rhon. E. Jones   :   http://www.beasleyallen.com/webfiles/
   Monsanto%20PCBs%20and%20Future%20Toxic%20Torts.pdf

(8) The Gadsden Times   :   Payments from PCBs fund being mailed to plaintiffs
:   http://www.gadsdentimes.com/article/20050723/NEWS/507230326

Conclusion

There are several different points that I would like to touch on in this conclusion to the book , General Electric and the Grand Experiment .

The first point would be : Are there , or were there , rather , any alternatives to the use of polychlorinated bipheyls (or any of it's 209 different derivatives , early on ? As early as the 1930s , perhaps ? Remembering , that as early as 1932 the dangers of PCBs to the health of the workers handling the substance were known . (1)

To quote : " Shortly after buying the 70-acre plant at the foot of Coldwater Mountain in 1935, the company learned that PCBs, in the doublenegative of one company memo, "cannot be considered non-toxic." A 1937 Harvard study was the first to find that prolonged exposure could cause liver damage and a rash called chloracne. Monsanto then hired the scientist who led the study as a consultant, and company memos began acknowledging the "systemic toxic effects" of Aroclors, the brand name for PCBs. Monsanto also began warning its industrial customers to protect their workers from Aroclors by requiring showers after every shift, providing them with clean work clothes every day and keeping fumes away from factory floors." (1)

" Now they know. They also know that for nearly 40 years, while producing the now-banned industrial coolants known as PCBs at a local factory, Monsanto Co. routinely discharged toxic waste into a west Anniston creek and dumped millions of pounds of PCBs into oozing open-pit landfills. And thousands of pages of Monsanto documents -- many emblazoned with warnings such as "CONFIDENTIAL: Read and Destroy" -- show that for decades, the corporate giant concealed what it did and what it knew. " (1)

" Monsanto enjoyed a lucrative four-decade monopoly on PCB production in the United States, and battled to protect that monopoly long after PCBs were confirmed as a global pollutant. "We can't afford to lose one dollar of business," one internal memo concluded." (1)

Instead , they ended up losing multiple millions - - - into the billions . $ 750 million in Anniston , Alabama alone .

The chemicals that are being used to replace polychlorinated biphenyl now are the following :

Use :

Use & Replacement	What PCBs were used for in the product
1) Transformers (electrical silicone and mineral oils machinery)	isolation fluid
2) Capacitors (electrical ester - based materials machinery)	isolation fluid
3) Sealants (construction) chlorinated alkanes	plasticizer
4) Paints and floor finishing chlorinated alkanes (construction design)	plasticizer
5) Carbonless Copy Paper isopropyl - substituted (office)	solvent for dyes
6) Electric cables halogenated organic compound and chlorinated alkanes	plasticizer
7) PVC antimony trioxide and aluminum trihydrate	flame retardant

above Source (2)

Now , these are compounds that are being used in modern times to replace PCBs .

But how about in the 1930's , when the dangers of PCBs were first becoming known ?

Well , there were several types of cooling systems available .

1) There was the ONAN system   : Oil system natural , air system natural   :
Simple . Hot oil flows to the top of the transformer , and is replaced by cool air on
the bottom . The hot air dissipated it's heat out into the open air ; and , as the oil
flow continued on in this manner , the transformer was cooled .

2) ONAF System   :   Fans (or ' forced cool air ') blowing on the top portion of
the transformer as the heat from the hot oil dissipates into the atmosphere .
Obviously , this speeds the process of cooling the hot il .

3) OFAF System   :   The cooling effect is accelerated by increasing the speed of
the flow of oil . Thus , a pump is placed near the bottom of the transformer to
accomplish just such a task .

4) OFWF System   :   Water being a better coolant than air ; water is placed at
the top of the transformer for the hot oil to dissipate it's heat into .

5) ODAF System   :   Cool air is forced , at pre - determined paths through which
the oil flows through ; alternating oil flow / air / oil flow / air / oil flow etc .

6) ODWF System   :   Works the same way that the ODAF system does ; except
that water is substituted for air . Source (3)

But , instead of these simple natural solutions that were available at the time ,
General Electric and Monsanto opted for polychlorinated biphenyls (PCBs)

Why ?

   Well , at the time PCBs were considered to be the ' ideal ' insulating fluid ,
because they were of a non-flammable nature and chemically stable . However , it
is just this ' chemical stability ' that now makes it so dangerous . In other words ,
it does not break down in nature . Once the health hazards of PCBs were being
exposed in the 1930s they could have been banned . I mean , a new product will
create problems until those problems are discovered , and then the product is
either fixed , or discarded . Monsanto and General Electric , however , allowed for
the continued use of polychlorinated biphenyls for between 45 and 50 years
before it was finally banned . A ' fix ' was never attempted .

   And how were PCBs actually banned ? It was not by a sudden ' burst of
conscience ' by Monsanto and General Electric . No ! Far from it ! In fact , General

Electric was caught when the clean water bill was enacted into law in the mid 1970s . Before then , there was just a tacit understanding between General Electric , and the various local governments that they were involved in (?) . I bring this point up only in regards to what I have heard when I lived up in New England (I'm now in Florida , U.S.) What I DID hear was that local government would either give permission , or turn a blind eye to the dumping of toxic materials ; in order to ' keep the wheels of industry turning ' ; create jobs etc . Looking at the towns of New England gives further credence to this theory . Some towns ; most towns ; remained rural , agricultural way stations , untouched by the (what turned out to be) ravages of the early industrial revolution . Some towns , however , such as Pittsfield , MA. , became industrial hubs because they had the correct confluence of rivers and streams that could provide water for the mills and factories to be able to turn . AND the town fathers would give a tacit nod for the industry to dump it's toxic materials . AGAIN , I do not wish to be-labor this point too greatly ; this is only what I have hard by word of mouth . And I do not stand and point the finger at Pittsfield city government back in the 1920s and 1930s .

 A case in point of this IS present , however , in the terribly polluted town of Anniston , Alabama :

 " By May 1970, PCBs were a hot topic in the national media. Members of Congress were calling for hearings. It seemed like only a matter of time before regulators would notice the river of PCBs spewing out of the Anniston plant. "This would shut us down depending on what plants or animals they choose to find harmed," the committee had warned.

 So Monsanto decided to inform the Alabama Water Improvement Commission (AWIC) on its own that PCBs were entering Snow Creek. And AWIC helped the company keep its toxic secrets.

 According to a company memo, AWIC's technical director, Joe Crockett, had been "totally unaware of published information concerning Aroclors (PCBs) ." The Monsanto executives assured him that everything was under control, and Crockett, who is now deceased, said he appreciated their forthright approach. "Give no statements or publications which would bring the situation to the public's attention," he told them, according to the memo.

 "In summary . . . the full cooperation of the AWIC on a confidential basis can be anticipated," the memo concluded. " (1)

This was the text of the AWIC's communication with Monsanto appointed an Aroclors Ad Hoc Committee , which had been formed in the 1970s to form a strategy to address the growing PCB problem . (10)

This would fit in exactly to the tactic that GE and Monsanto are using at present ; that of re-locating to areas (Eastern Europe , China , Southeast Asia) that allow it to dispose of it's toxic waste . The present day pollution of China speaks for itself .

As the problems became worse in the early 1970s : Before the year was over, Crockett helped out once more. The Justice Department was considering a lawsuit against Monsanto over PCBs, and the EPA wanted it to dredge Snow Creek. So Crockett set up a meeting between Monsanto and an EPA regulator and helped argue the company's case. The company's problems disappeared. One executive noted with relief in a memo that a federal prosecutor had tried but failed to obtain Monsanto's customer list: "I shudder to think how easily it would have been for someone . . . to start spilling the beans as to whom we have been selling PCB products." (1)

So , government , Monsanto and General Electric managed to sweep the problem under the rug ...... for a while .

Until 1975 .

In 1975 , General Electric had long since known of the health problems . And had done nothing about it . It was only when the public outcry grew extremely loud over the PCBs and pollutants were found in the Hudson River , in New York U.S. , that permits were now required , further government investigations were launched ; and the dumping of PCBs was finally prohibited between 1977 and 1979 .

Not every problem was immediately taken care of , or addressed , however .

In Anniston , Alabama " Monsanto's luck with regulators held in 1983, when the federal Soil Conservation Service found PCBs in Choccolocco Creek, but took no action. In 1985, state authorities found PCB-tainted soils around Snow Creek, but a dispute over cleanup details lingered until a new attorney general named Donald Siegelman took office in 1988. In a letter that April, Monsanto's Anniston superintendent thanked Siegelman -- who is now the state's Democratic governor -- for addressing the Alabama Chemical Association, and meeting Monsanto's lobbyists for dinner. Then he got to the point: Monsanto wanted to go forward with its own cleanup plan, dredging just a few hundred yards of Snow Creek and its tributaries." (1)

It was the same in Pittsfield , MA,. , as disputes and delays pushed the start of the clean - up into the 1990s and early 2000 . In 1998 , when we sold our house in Pittsfield , MA (which was directly across the street from the General Electric , and just down the road from Silver Lake) , this scandal and debate was just breaking amongst the local populace . We were not there for any subsequent soil testing on the property , which I heard was extensively done on many properties in the Pittsfield area .

But now let's get off of the sordid details of the way in which GE and Monsanto were caught dumping their toxic waste . A deed that they hid from the public for YEARS , simply for profit ... or so it would seem .

But was profit the only motive ?

I wish to look at the ' toxicolgy testing ' issue just a bit .

What is toxicolgy testing ? What does it involve ?

Web MD : " A toxicology test checks blood, urine, or saliva for the presence of drugs or chemicals. In rare cases, stomach contents or sweat may also be checked." (4)

Well , this is a good thing , right ? Checking for the presence of toxic chemicals in the human bloodstream , or the urine or saliva ?

But .... if it is used for a different purpose ?

General Electric and Monsanto let the poisoning of their workers go on for years , after they and others knew of the harmful effects of polychlorinated biphenyls . There was a profit motive , no doubt .

But .... was this the same type of toxicolgy testing and experimentation that went on in the nazi concentration camps during the years 1935 / 1936 through 1945 ? This is approximately the same time period that the toxic chemical pcb was allowed to be dumped on workers in GE's factories , and also dumped on nearby land , and into water supplies . And continued to be dumped even after the toxic effects were known .

Let's look at toxicolgy tests during the war (World War 2) The most famous of these tests took place in China , after the Japanese invaded in 1932 . Wikipedia : " In Japan, Unit 731, located near Harbin (Manchukuo), experimented with prisoner vivisection, dismemberment and induced epidemics on a very large scale from 1932 onward through the Second Sino-Japanese war. With the expansion of the empire during World War II, many other units were implemented in conquered cities such as Nanking (Unit 1644), Beijing (Unit 1855), Guangzhou (Unit 8604) and Singapore (Unit 9420). After the war, Supreme commander of occupation Douglas MacArthur gave immunity in the name of the United States to all members of the units in exchange for a tiny part of the results, so that in post-war Japan, Shiro Ishii and others continued to hold honoured positions. The United States blocked Soviet access to this information. However, some unit members were judged by the Soviets during the Khabarovsk War Crime Trials. The effects were lasting and China is still working to counteract the effects of buried pathogen caches." (5)

The United States gave pardon to these criminals for a small sampling of the results .

In the nazi / German camps , again from Wikipedia :

" The Herero and Namaqua Genocide in present day Namibia, in Southern Africa, resulted in a large number of prisoners in Nazi concentration camps. These prisoners were used as medical test subjects by German agents.]

During the second World War, Nazi human experimentation occurred in Germany with particular bias towards euthanasia. At the war's conclusion, 23 Nazi doctors and scientists were tried for the murder of concentration camp inmates who were used as research subjects. Of the 23 professionals tried at Nuremberg, 15 were convicted. Seven of them were condemned to death by hanging and eight received prison sentences from 10 years to life. Eight professionals were acquitted. (Mitscherlich 1992)

The result of these proceedings was the Nuremberg Code. It includes the following guidelines, among others, for researchers:

Informed consent is essential.
Research should be based on prior animal work.
The risks should be justified by the anticipated benefits.
Research must be conducted by qualified scientists.
Physical and mental suffering must be avoided.
Research in which death or disabling injury is expected should not be conducted." (5)

More om German ' meduica; experimentation ' , again from Wikipedia :

" Prisoners were coerced into participating; they did not willingly volunteer and there was never informed consent. Typically, the experiments resulted in death, disfigurement or permanent disability, and as such are considered as examples of medical torture." (6)

This relates to the toxicology ' tests ' done by GE and Monsanto , in that they left many , many people disabled or disfigured (or dead) in the footprint of where GE and Monsanto had been .

But were these , in actuality , toxicolgy tests that were being conducted by General Electric and Monsanto ?

Well , personally , I know that I have been studied countless times for my type 1 diabetes ; prodded and poked by the medical profession ; given all type of drugs ..... most of which did no good at all .... or , they have a negative effect . Now , multiply my experience with thousands of others in the Pittsfield area , in the Anniston , Alabama area ; or the Schenectady , New York area ; or along the Hudson river in New York State , where GE dumped an estimated 1.3 million pounds of PCBs between 1947 and 1977 (7) Yes , multiply my case by many cases .

But , what are they studying ? What is the purpose of all of this endless testing and data ?

For the nazis and the Japanese , it was to find out how much poison the prisoner could stand before he (or she) died , or was disabled . Of course , it is well known that the nazis were trying to build a ' super race ' . But , how could they build a ' super race ' when they knew about the poisons that they had spewed across Europe . And ... knew about the poisoning of humanity that would take place in the future .

That future belonged to Monsanto . After World War 2 , Monsanto , GE and friends went into high gear ; and began the addition of GMOs to our food , and also the massive use of chemicals designed to kill insects and weeds . I am referring here , in part , to the weed killer ' round-up ' , or ' Glyphosate ; a herbicide that was very widely used by farmers in the United States . The toxic effects of Glysophate (roundup) sprayed food is still being debated , but the herbicide is banned in California , U.S. (8)

Like I mentioned earlier , the use of glysophate is still being debated in some areas - - - we have yet to know the long term effects of being exposed to this widely used herbicide .

But looking at the larger picture ; what did Germany and Japan want to learn from their toxicity / toxicolgy tests ? To build a ' super race ' . A super race that is capable of surviving in a highly polluted world , apparently . Monsanto and GE , as I mentioned earlier , are now highly involved in the trans - humanistic movement (I devoted a chapter to this earlier) . And transhunism is a movement that is dedicated to moving ' beyond ' where humanity is at this present moment . And this consists of artificial intelligence ; a large usage of robots and robotics ; in fact the combination of the human (even if it is only the human mind) with the robotic body .

But what type of world is this ' robot - man ' going to live in ? It appears as though the ' perfect ' Monsanto and GE world would consist of polluted air , polluted water , highly chemicalized food , blackened lakes and rivers ...... in very fact , a literal ' hell on earth ' . This coincides with what we used to call certain sections of the General Electric in Pittsfield - - - ' Hell on Earth ' . I did not , at the time , know the exact locations of these sections .... word had just filtered down to me from talk around town that this is what they were called . Now , I am just assuming that some of these places were the PCB pits .

Not a pretty picture - - - a blackened ' hell ion earth ' , populated by robots ; controlled by a small number of ' controllers . And this is their picture of perfection - - - when they already had a perfect world around them . Pittsfield , Massachussetts is a beautiful area ; or WAS a beautiful area .... before GE moved in . The surrounding areas consist of beautiful farmland , rolling hills , dairy farming , huge apple orchards . The foliage and the mountains were / are just exquisite . Who would want to change that ? GE sure would . They took these honest , hard working fellows from this area , including my grandfather , Joseph Taikowski , Sr . , and put them to work in the ' pits of hell ' . And this is the future that Monsanto and GE envision for all of us . Life in the ' pits of hell ' ; controlled by the very few , who are supposed to benefit from the knowledge derived from these toxicology experiments . As these experiments were not performed to benefit the mass of humanity . Oh , no ; no way . Perish that thought ! Only the very few - - - the ' elite ' . Those who would control this poisoned world , and be able to survive with the superior knowledge of what was going on , and the superior ' super ' medications that they developed to deal with the sicknesses that developed from the poisoning .

This is the future as envisioned by GE and Monsanto . We are all lab rats in the experiment ; until such time as we are no longer needed ; then we are eliminated . The workers that built this poison system are the first to go , having been exposed first hand to the toxic effects ..... but their families are cursed also , with many of these sicknesses being passed down through generations through damage to the genetic structure ; in diseases like diabetes , where the genes tell the immune system to destroy different parts of the body , instead of protecting it against invasive germs , etc . All about communication in the genetic structure . That is why the work of Russian physicist and molecular biologist Pjotr Garjajev and his colleagues in Russia is so important . To re-establish this wondrous communication between the cells - - - where it has been so rudely interrupted . And , by extension , to re-establish the communication between human beings , which mimics this study ; the communication which allows us to know , and to feel what is going on around us . Do we realize this now ? Some do ; most just talk on cell phones , in an attempt to communicate properly . But this attempt fails , more often than not .

Yes , GE and Monsanto inherited a good world - - - a rural world of farms and builders . And they destroyed it . For their image of a machine controlled society . And this is the future that we are heading into .

But ; perhaps Monsanto and General Electric have learned a lesson from all the sanctions and penalties that they have incurred - - up to $ 750 million in Pittsfield , $ 750 million in Anniston , Alabama , and over $ 1.5 billion to clean up the Hudson River .

I'd like to quote once again from ' GE Miisdeeds ' ; a watchdog group that is dedicated to keeping an eye on General Electric :

" What distinguishes General Electric is not merely the number of crimes committed -- or the dollar amount of the crimes -- but a consistent pattern of violating criminal and civil laws over many years. Even worse, General Electric has been a leader in using political influence to attempt to overturn environmental and defense contracting laws that GE persistently violates. (9)

They never learn . They just move the operation to another country that does not have stringent environmental laws .

GE has repeatedly denied that PCBs are a health hazard , in spite of all the evidence to the contrary .

From 1999 :

" Another flank of GE's strategy is to challenge the conventional wisdom that PCBs are all that toxic to begin with. "There is no credible evidence that PCBs cause cancer" GE wrote in a 1999 report, a line company officials including Jack Welch, have repeated since. " (10) Jack Welch , by the way was CEO of General Electric from 1981 to 2001. General Electric has also claimed that the workers at it's plants had little to no exposure to PCBs . (10) Which is , of course , an out and out lie . In a round-a-bout argument , GE has also hired doctors who say that PCBs do not cause cancer (Dr. David Carpenter is one of these doctors) ; when " Every international group of experts that has been asked to look at the issue has concluded that they are proven to cause cancer in animals and are probable carcinogens in humans. " (10) Odd that GE would hire this team when they claim that their workers were never exposed to PCBs.

What is happening here , with these oddball statements , is that GE is attempting to win in the court of public opinion by sewing doubt ; even while ignoring the scientific facts .

Jack Welch , GE's gregarious CEO for twenty years , even had this to say about PCBs : " We don't believe there are any significant health effects from PCBs." ; and , about the clean up efforts : " Let me just tell you, as I tried to tell you in my report, we use sound scientific principles, we move forward and clean up past legal issues and we have no qualms at all about spending the right amount of money to get it done. To throw money at subjects that do not require it makes no sense."'   (11)

In spite of Mr. Jack Welch's objections " , in 1976, Environmental Conservation Commissioner Ogden Reid charged GE with violating water standards in the Hudson River. During the trial, Dr. Gerald Lauer, one of GE's expert witnesses, testified that the fish caught near the electrical plants showed no contamination above FDA limits." (12)
" The next day, however, an attorney from the Natural Resources Defense Council (NRDC) forced Lauer to admit that decimal points had been shifted in his reports and in fact all of the fish were highly contaminated." (12) (Note on this : the previous day , in testimony , an ' expert ' from GE had testified that the fish were only slightly contaminated (12)

" Judge Sofaer found GE liable under the 1972 Clean Water Act for "corporate abuse" of its state-issued permits." (12)

As I noted before , the cleanup of the Hudson River will cost GE approximately $ 1.5 billion dollars . (The on-going cleanup , I might add)

So , is GE getting it's just deserts ? Or ... are they simply engaging in the same tactics as they did in the years 1930 through 1977 ; when they dumped toxins , at will , with no restraint , at multiple plants here in the United States . No , it's obvious that they have not learned a thing from these massive fines . They have merely moved their operation overseas , where the process begins anew .

Here in the U.S. we can only hope that the younger generation benefits from the eviction notices that GE received . Of course , PCBs stay in the environment for years , so there is still reason for concern .

We are engaged in a battle to save the Earth from the diseased , darkened world that GE envisions as our future . They certainly did their level best to spread disease far and wide . If the Earth were given the chance , it might show the management lackeys a thing or two about the wonders of communication , the wonders of creation . But GE , in it's fear , chose to view the Earth as an adversary ; and now we all must pay the price of their failed vision - - - the failure of ' The Grand Experiment ' .

(1) Organic Consumers Association : Monsanto Hid PCB Pollution for Decades : https://www.organicconsumers.org/ old_articles/monsanto/pcbs010702.php

(2) Chemistry12-pcb : Alternatives to PCBs : Replacement chemicals : http://chemistry12-pcb.blogspot.com/

(3) Electrical4U : Transformer Cooling System and Methods : http://electrical4u.com/transformer-cooling-system-and-methods/

(4) Web MD : Toxicolgy Tests : http://www.webmd.com/a-to-z-guides/toxicology-tests

(5) Wikipedia : Experimentation on prisoners : http://en.wikipedia.org/wiki/Experimentation_on_prisoners

(6) Wikipedia : Nazi human experimentation : http://en.wikipedia.org/wiki/Nazi_human_experimentation

(7) Riverkeeper : Hudson River PCBs : http://www.riverkeeper.org/campaigns/stop-polluters/pcbs/

(8) Landscape Online : California Herbicide Ban Causes Potholes, Overgrown Weeds : http://www.landscapeonline.com/research/article/8469

(9) Clean Up GE : GE Misdeeds : Conclusion : http://cleanupge.org/gemisdeeds.html

(10) Clean Up GE : Toxics on the Hudson : http://www.cleanupge.org/pcbarticle.html

(11) Clean Up GE : You Don't Know Jack ! : http://www.cleanupge.org/jacksays.html

Image Credits